Urban Sustainability

The Urban Sustainability Book Series is a valuable resource for sustainability and urban-related education and research. It offers an inter-disciplinary platform covering all four areas of practice, policy, education, research, and their nexus. The publications in this series are related to critical areas of sustainability, urban studies, planning, and urban geography.

This book series aims to put together cutting-edge research findings linked to the overarching field of urban sustainability. The scope and nature of the topic are broad and interdisciplinary and bring together various associated disciplines from sustainable development, environmental sciences, urbanism, etc. With many advanced research findings in the field, there is a need to put together various discussions and contributions on specific sustainability fields, covering a good range of topics on sustainable development, sustainable urbanism, and urban sustainability. Despite the broad range of issues, we note the importance of practical and policy-oriented directions, extending the literature and directions and pathways towards achieving urban sustainability.

The series will appeal to urbanists, geographers, planners, engineers, architects, governmental authorities, policymakers, researchers of all levels, and to all of those interested in a wide-ranging overview of urban sustainability and its associated fields. The series includes monographs and edited volumes, covering a range of topics under the urban sustainability topic, which can also be used for teaching materials.

Xiaohui Hu

Uneven Economic Resilience of Old Industrial Cities in China

A Multiple-Perspective Analysis

Springer

Xiaohui Hu
School of Geography, Jiangsu Center for Collaborative Innovation in Geographical Information Resource Development and Application
Nanjing Normal University
Nanjing, Jiangsu, China

ISSN 2731-6483 ISSN 2731-6491 (electronic)
Urban Sustainability
ISBN 978-981-99-9278-2 ISBN 978-981-99-9279-9 (eBook)
https://doi.org/10.1007/978-981-99-9279-9

This Springer imprint is published by the registered company Springer Nature Singapore Pte Ltd.
The registered company address is: 152 Beach Road, #21-01/04 Gateway East, Singapore 189721, Singapore

This book is dedicated to Chanjuan and Jiachen, without whose love, company and support this book would have been an impossibility;

It is also a proof of my own academic journey literally starting from 2011. The book teaches me to being resilient when facing any crises

Foreword

The author of this book earned his Ph.D. in Economic Geography from Kiel University, Germany, in 2015. With a long-standing dedication to research in economic geography and regional development, a significant portion of the content within this book emanates from a continuous accumulation of publication achievements since his doctoral graduation up to date. These achievements well focus on the restructuring progress in old industrial cities and regional economic resilience. The book thus showcases both the originality of scholarly research and a profound level of expertise.

This systematic study unfolds the processes, impact factors and their underlying mechanisms that give rise to the unevenness of regional economic resilience across Chinese old industrial cities (OICs). It leverages a rigorous theoretical framework that substantiates the core arguments. The book uniquely amalgamates theoretical constructs with comparative case studies, harmonizing quantitative and qualitative methodologies to offer fresh empirical evidence. The proposed call for adopting multiple but also integrated perspectives from EEG, IEG, REG and agency theories, namely an 'engaged pluralism' approach, is essential for future's research on the topic for the restructuring of Chinese OICs.

Undoubtedly, this book stands as a notable and distinctive contribution to the discourse on regional economic resilience within China and beyond. Scholars and practitioners in the fields of and/or with the backgrounds of economic geography, urban and rural planning, urban studies, and public administration, are all potential readers. This work is poised to captivate the interests of both undergraduate and postgraduate students, as well as practitioners, researchers and policy-makers seeking an advanced understanding of uneven regional economic resilience in an increasingly uncertain and risky world.

September 2023

Prof. Canfei He
Peking University
Beijing, China

Preface

My book is concerned with the uneven economic resilience of old industrial cities in an evolutionary and institutional, as well as a human agency perspective. It also proposes a set of conceptual perspectives that can be integrative and fitting for doing regional economic resilience research in the increasingly uncertain world, particularly concerning the foreseeable long-term impact from both an emerging and broader global shift of geo-political economies and the COVID-19 pandemic crisis.

There is considerable literature in economic geography (EG), focusing on the role of firms and industry dynamics in affecting the evolution of old industrial cities (OICs). However, little work has been done, both empirically and theoretically, on the role of state, multi-scalar institutions and relations, politics and policies. Moreover, the majority of the research on the topic is very much based on single-case studies, whose evidence and knowledge are mainly derived from European and American contexts, and are centred on resilience under economic/financial crises (particularly the 2008 global financial crisis). Given these research gaps, the key aim of the book is to (1) theoretically, integrate an institutional, relational and geographical political economy perspective with the Evolutionary Economic Geography (EEG) approach, in order to give a better understanding of the nature and its logics of the unevenness of economic resilience in OICs, and (2) empirically, to deliver a comparative setting of both quantitative and qualitative research, by closely examining what impact factors affect regional economic resilience and how (also to what extent) different Chinese OICs adapt to different sorts of crises, such as local exhaustion of nature resources, the 2008 global financial crisis, as well as the COVID-19 pandemic crisis.

My book pays more attention to China's specific institutions and institutional-social contextual realities, particularly giving more focus on how the state, more precisely, how human agency within the state and state-owned enterprises (SOEs) of OICs, responds to and enacts different types of shocks and crises for long-run industrial adaptation and adaptability. It does not simply focus on the micro behaviors of firms, states and people, but more on their embedded positions and relationships of multi-scalar political-institutional frameworks and policy fields. The empirical research is not only based on the use of quantitative modeling and regression methods to identify the determinants of regional economic resilience in Chinese resource-based cities facing the conventional 2008 global economic crisis and the COVID-19 crisis, respectively. It is also based on in-depth case studies on two Chinese coal-mining cities (Zaozhuang in Shandong Province and Fuxin in Liaoning Province) both facing the 'slow-burn' crisis of local coal resource exhaustion since 2000.

In this book, the concepts of path dependence, lock-ins and industry path development are critically examined—by adopting and integrating perspectives of IEG, REG and agency. The empirical evidence suggests that the effects of lock-ins in Chinese contexts are more politically constituted, due to the Chinese-specific administrative hierarchy system among SOEs and governments. Despite strong industrial path dependence and negative regional lock-ins, new paths are not restricted to emerge. Based on the notion of path plasticity, I conclude that the Chinese local state and its state leaders in the forms of state governance, institutional arrangement and policing in OICs play a key role in affecting the ways in which adaptation and adaptability take place and interact in regional economic resilience. The case studies also well address how structure and agency factors, actors and forces are intentionally organized to work and to respond to shocks/crises. Chapters 6 and 7 deliver a key idea that state agencies are aware of how to balance the interests among multi-scalar political economies, and of how to flexibly read and enact locally according to multi-scalar institutional and policy conditions. These actions lead to divergent processes, mechanisms and outcomes of regional economic resilience.

The most important contribution of the book is the comparative analyses in which regional economic resilience is examined and understood in three main different crises. This contribution goes beyond the EEG-based mainstream understanding of regional economic resilience with either a local structure-centred approach or an agency-based perspective. By giving more emphasis on the roles of multi-scalarity, contexts and the nature of the crisis itself, this book reconceptualizes regional economic resilience, with a comparative set of empirical evidence from the non-Western experience provided. In this book, a more advanced conceptual framework with *multiple perspectives* on analyzing and understanding the uneven economic resilience of OICs and beyond has been developed with three dimensions:

First, it is built upon an in-depth conceptualization that de-constructs the dualism idea between adaptation and adaptability. And it redefines them in an interactive, dialectical and evolutionary way.

Second, this book highlights the importance of multi-scalar institutions and national political economy change in coordinating the behaviors of firms, local states and the micro dynamics of their constitutive human agents in China's OICs. The historically conditioned political distance between localities and the central state is essential. Therefore, adopting an IEG, GPE, and agency combined perspectives can help to fundamentally understand the differentiated economic resilience of China's OICs.

Last but not least, concerning the on-going crisis of the COVID-19 pandemic and its possible mutations, the book suggests a relational perspective of regional economic resilience by combining notions of supply chain resilience, industry chain resilience and global value chain resilience. This perspective is particularly useful to understand the economic resilience of OICs on the ground, if (most likely, indeed) those economies are well embedded in, and linked to global, national and any cross-regional industry systems. Such relational view and the notion of chain resilience can serve as an important complement to the EEG resilience conceptualization, which is proofed as a powerful explanatory tool for studying regional economic resilience in China. In the end, through this book, I argue that, in an increasingly uncertain and globalized "connected" world, harnessing multiple perspectives of regional economic resilience should be placed at the core to explore the question of why some cities in China are able to navigate and form strong and sustained resilience, while others fail to do such with weakened resilience.

Nanjing, China Xiaohui Hu

Acknowledgements I would like to express my gratitude to all the co-authors of the papers that constitute this book. Thank you for their support and consent for republishing. This work is supported by the National Social Science Fund of China (Project No. 22JLB00738).

Contents

Abbreviations

COVID-19	Coronavirus disease 2019
EEG	Evolutionary Economic Geography
EG	Economic Geography
FDRC	Fuxin Development and Reform Commission
GPE	Geographical Political Economy
GPN	Global Production Network
GSC	Global Supply Chain
GVC	Global Value China
IEG	Institutional Economic Geography
NDRC	National Development and Reform Commission
OICs	Old Industrial Cities
RECs	Resource-Exhausted Cities
REG	Relational Economic Geography
SOEs	State-Owned Enterprises
TVEs	Township and Village Enterprises
VOC	Varieties of Capitalism
ZDRC	Zaozhuang Development and Reform Commission

Chapter 1
Theories, Concepts, and Research Accounts of Regional Economic Resilience

1.1 EEG, Adaptation, Adaptability and Resilience: Problems and Limitations

In recent years, the term of 'adaptation' and 'adaptability' has become a conceptual fashion in the EG literature (Boschma, 2015; Hassink, 2010a; Martin, 2012a, 2012b; Pike et al., 2010). Along with the 'resilience' concept, these terms have been increasingly engaged with the analysis of how regional economies response and adapt in face of varieties of shocks and disturbances in a post-crisis, globalized and uncertain context (see two special issues: 'The Resilient Region', 2010, 3(1) in Cambridge Journal of Regions, Economy and Society, and recently, 'Recession, Recovery and Resilience 2014, 48(11) in Regional Studies). With the evolutionary 'turn' in EG in general, and also with the policy agendas for sub-national economic recovery in the wake of the 2008 global crisis and the 2019 COVID-19 pandemic crisis in particular, evolutionary economic geographers have paid much attention to the notion of resilience, adaptation and adaptability in understanding the geographical differentiation of response and redevelopment of regions (Boschma, 2015; Martin & Sunley, 2015; Pike et al., 2010). In recently, The COVID 19 pandemic weakened cities' conventional advantages such as economic variety, skilled labor force, agglomeration, and economies of scale. The persistence of evolving and mutating coronaviruses suggests that research on regional economic resilience is crucial and needs to take the nature of external shocks seriously and pay more attention to the role of geographies and contexts in understanding resilience (Glaeser, 2022; Martin & Sunley, 2020).

The question of "why do regions differ in resilience", inquired by Martin and Sunley (2015, p. 25), is at the core, but an under-explored, question in (E)EG. Although the resilience concept itself bears inherent deficits, primarily concerning its '*rebound back*' (multi-) equilibrium thinking, it does shed some light on the uneven mechanisms of long-run regional development, by explaining differences in how regions react to a succession of shocks and challenges (Martin & Sunley,

X. Hu, *Uneven Economic Resilience of Old Industrial Cities in China*,
Urban Sustainability, https://doi.org/10.1007/978-981-99-9279-9_1

2015, p. 37). With this respect, resilience can be regarded as contextually sensitive and dynamic capabilities of regional economies. It ought to be understood not only as short-term recoverability and resourcefulness to adapt to contextual changes, but also as long-run transformability and sustainability for system-level building of institutions and future development.

In the light of Grabher and Stark (1997)'s early work, Pike et al. (2010) claimed "adaptation and adaptability as central causal concepts in explaining the geographically uneven resilience of places" (p. 60). In general, adaptation is considered as an ability of changes within existing structures and paths in order to fit/adapt to the changing environment. It is, to some extent, similar to the notion of 'adjustment' featured by slow, incremental and path-dependent development (Grabher, 1993; Chapman et al., 2004). Adaptability, however, is more subject to a capability of firms, people and states purposively seeking for alternative developmental options, and being dedicated to innovation and structural change. It can be largely associated with the so-called 'reorientation' process in regional economies that has been claimed by leading scholars (Boschma, 2015; Grabher & Stark, 1997; Martin, 2012a).

The existing accounts on adaptation and adaptability relating to regional economic uneven evolution have tended to draw a divide between the two. That is, there is a *trade-off* conceptualization between adaptation and adaptability. It has been argued that adaptation (characterized by tightly connected agents primarily focusing on preconceived paths/industries) basically undermines adaptability (featured by loosely tied agents endeavoring on path creation/renewal) (Grabher, 1993; Pike et al., 2010). Martin and Sunley (2006)'s seminal work on path dependence and regional economic evolution, again, highlights this idea, by saying "there is a trade-off between specialization and a short-lived burst of fast regional growth on the one hand, and diversity and continual regional adaptability on the other" (p. 421).

However, in this book, I argue that this trade-off idea of adaptation and adaptability is an inappropriate, dualistic and reductionist conceptualization. Because, in reality, stakeholders and agents in the economy often act in a heterogeneous way and at different scales, and harness/mobilize different levels/power of assets. These actions are embedded in, and affected by, incoherent power relations and differential aims and motives, given overlapped economic, social and political interests and layers (Ibert & Schmidt, 2014). It is thus possible that adaptation and adaptability co-exist, and even compete in a certain region (see details in Chap. 8). Moreover, Boschma (2015) stresses that in an evolutionary view regional economic resilience can be conceptualized as a long-term ability to overcome the trade-off between adaptation and adaptability.

I take this point seriously, as EEG, in general, and the resilience notion, in particular, has suffered from a lack of critical rigor and conceptual clarity. They seem to have been in a black box with clear conceptualizations, becoming a sort of 'fuzzy concepts', both in academia and policy fields (Christopherson et al., 2010; MacKinnon & Derickson, 2013; Markusen, 2003). As stated by Martin and Sunley (2015), no matter single-case or comparative studies on geographies of regional resilience, "…without some agreed and meaningful concepts we have no way of assessing whether and to what extent regional economies are resilient" (p. 36).

Thus, the starting point of departure of this book is to carefully examine the yet highly contested concept of regional economic resilience itself. This book particularly focuses on the notion of adaptation and adaptability. It aims to understand their explanations in the understanding of regionally differentiated economic resilience in an increasingly uncertain context facing different types of crises. In this book, the term 'adaptability' is adopted as a keyword in this book. Because, I do not merely view it as an ability to create regional new growth paths. But, beyond it, adaptability here refers to a dynamic and evolutionary ability of process that is internally and logically linked to, mutually inclusive with, rather than separated from 'adaptation'. As such, adaptability is perceived as an integral part of potentials for long-term regional economic development—as the key of understanding regional economic resilienc in EEG.

The primary objective of this book is to systematically address three pivotal conceptual research inquiries:

(1) How can the concepts of adaptation and adaptability be more precisely delineated within in a regional context?
(2) What are the influential factors that determine the nature and structure of regional economic adaptation, adaptability and their combinational understanding of resilience?
(3) In what manner and for what reasons do diverse regional economies exhibit varying patterns of adaptation over an extended temporal framework, consequently giving rise to disparate geographies of economic resilience?

1.2 Why Economic Resilience of Old Industrial Cities Matters: Navigating The Missing Link

In existing literature on regional adaptation, adaptability and resilience, old industrial cities (from here OICs) have drawn considerable attention by economic geographers. Because OICs often confront a dilemma of coping with the problem of being locked in old existing mature industrial specializations, and seeking for new resources and opportunities toward economic renewal and diversification. Besides, they are also economically vulnerable and more inclined to be exposed to external shocks, and/ or more likely to suffer from internal and structural crises. Therefore, economic resilience, namely, the methodology, agency and capability of adaptation, matter significantly for OICs.

Much work on OICs has been done on the role of path dependence and lock-ins in constraining the restructuring process. It mainly emphasizes rigid firm-based *organizational* inter-relations and subsequent *cognitive* and *political* factors protecting existing structures (Eich-Born & Hassink, 2005; Grabher, 1993). However, evidence has proven that path dependence does not necessarily lead to lock-in. And lock-in itself does not necessarily refer to path stability and continuity as well. Hassink (2007) well articulates that OICs with (relatively) 'weak ties' of local actors have 'strengths' to avoid lock-ins. Such kind of tie structures can lead to high adaptability for future new industry formation. In this respect, lock-in can be conceptualized as a process of change instead of static stage. It means that under certain conditions lock-ins can turn out to be possibilities or sources for regional renewal and path creation (Martin, 2010, 2012b). Many studies have also shown that new paths do not emerge from scratch, but tend to draw assets from existing competences, networks, and structures (Martin & Simmie, 2008; Martin & Sunley, 2006; Neffke et al., 2011). In other words, adaptation (affected by the effect of path dependence) can lead to or mutate into a stage that facilitates adaptability, through the process of '*related variety*' (Frenken et al., 2007) or '*regional branching*' (Boschma, 2011) enabled by local actors' entrepreneurial actions (Garud et al., 2010; Malecki, 2009).

The co-existence of adaptation and adaptability in OICs has been emphasized, albeit in an implicit sense. Strambach and Halkier (2013), for instance, state that development paths in regions are not coherent in themselves, and there is, always, a degree of *plasticity.* This means that even in a well-established institutional setting there remains leeway for actors to adopt strategic actions. New growth paths can be fostered without a need of breaking the existing paths/structures. Drawing upon evidence in Wales, UK, Morgan (2012)'s work, for instance, shows that under the context of particular politics and state interventions, local actors were not merely involved in adaptation relating to path extension or persistence. They, also, at the same time, tried to cultivate adaptability by fashioning new paths. These studies show that adaptation and adaptability can co-exist in parallel. Adaptation may be not at the expense of adaptability and vice versa. Besides these studies, evidence shows that the nature of adaptability and its relation with adaptation vary from one OIC to another. They tend to be more related to the role of social agency, institutions, politics and policies, rather than to traditional forces of firms and industries (Dawley, 2013; Christopherson et al., 2014; Hu, 2014; Hu et al., 2022; Hudson, 2005; Isaksen, 2014; Tan et al., 2020).

However, by far, the existing literature on the restructuring of OICs has been relatively weak on how adaptation and adaptability are defined in a dynamic and interactive perspective. In general, three research gaps can be identified: First and foremost, the human agency perspective is insufficiently integrated into the concept of path dependence, lock-in and path creation in the OIC context. We are still not clear about how different actors, particularly key individuals or groups play a role in enabling or hindering adaptation and adaptability, and who they are? An in-depth investigation on the role of entrepreneurship and place leadership might be potentially useful to articulate the dynamics of adaptability, as well as the inter-relationship concerning to adaptation between institutions, politics and policies.

Secondly, too much work has been done on the role of local endogenous firm-based forces in affecting the restructuring of OICs. However, multi-scalar and system-level resources, agents, agency that interactively work together on the (re)organization of adaptation and the possibilities of adaptability are far from well explored. Last but not least, the role of institutions and related broader political economies in the adaptability of OICs is under researched. In particular, the role of the state, institutions and policy at several spatial levels should be placed as a crucial point in understanding the dynamics and the interplay between adaptation and adaptability (Bristow &

Healy, 2014). It is the very case in the context of OICs. As local economic actors in OICs are relatively weak in promoting new paths, the state may play an active role in introducing external forces for regional industrial change.

Based on above discussions, the existing research weaknesses and gaps on regional economic resilience need to be coped with by a systematic and comparative in-depth research. As Boschma and Martin (2010) noted that "to understand the nature of uneven evolution of regions, we need to first understand how the processes of path creation and path dependence interact to shape geographies of economic development and transformation" (p. 7). This book thus seeks to shed some light on the understanding of regional uneven economic resilience, by a particular focus on the interactions and relations between adaptation and adaptability in a regionally comparative perspective.

Besides, the existing studies on regional economic resilience of OICs have not taken the nature of crisis and/shocks seriously. Most of them often tacitly place the examination of regional economic resilience within a broader macroscopic framework encompassing external perturbations, such as the 2008 financial crisis or other economic catalysts. Martin et al. (2016) argue that the scale, nature, duration of a shock/crisis matter siginificantly for the extent and ability of an economy recovering from and responding to. In some extreme cases, the nature, type and scale of crisis can fundamentally determine the process, ability and directionality of regional economic resilience (Hu et al., 2022).

Moving from the 2008 global financial crisis, the 2019 coronavirus disease (COVID-19) pandemic, as an on-going global pandemic continually spreading across the world, has led to a truly worldwide crisis. Since COVID-19 spreads between people who are in close physical contact and geographical proximity, human activities of any kind as long as requiring mass gatherings are strictly restricted (Hu et al., 2022). Only for months, the pandemic has led to the largest global economic disruption since the Great Depression (Nicola et al., 2020). Evidence also strongly shows that OICs suffered more negative impact from COVID-19 and confronted with more severe unemployment, economic decline and population shrinkage, which are above the levels experienced in the wake of the 2008 financial crisis. Given such an unprecedented scale and impact depth of the crisis, the resilience idea has received a renewed deal of research attention.

Hence, the second fundamental objective of this book is to earnestly consider the backdrop of the COVID-19 pandemic crisis and to incorporate its inherent characteristics into the conceptual frameworks and comprehension of regional economic resilience. It is evident that the discourse pertaining to how regional economies react to and navigate through the crisis, a facet closely tied to regional economic resilience, has not yet assumed a prominent role within the COVID-19 literature (OECD, 2020). This brings forth several inquiries that warrant heightened scrutiny within both academic and policy domains:

(1) How do regional and urban economies withstand and react to the challenges posed by COVID-19?

(2) What factors exert substantial influence over the spatial patterns of regional economic resilience amid the COVID-19 context, particularly concerning economically disadvantaged regions such as OICs?
(3) What form of pragmatic policy implications, if discernible, can be harnessed to bolster regional resilience against the likes of COVID-19 and analogous crises in the foreseeable future?

1.3 Key Research Questions and Rationales

The book thus aims to advance the understanding of the geographies of regional economic resilience of OICs facing differentiated nature of crises on the ground. By acknowledging the potential limitations of using single explanatory concepts or even certain paradigms (Hassink et al., 2014; MacKinnon et al., 2009), this book synthesizes several evolutionary, relational, institutional and human agency-sensitive approaches in investigating the geographies of regional economic resilience of OICs in the Chinese context. Moreover, although the primary research aim is not to 'theorize back' on EG based on Chinese experiences (Yeung & Lin, 2003), I do realize the importance and specificities of the Chinese context, and the possible extreme practices and actions relating to regional resilience on the ground that might exclude from the mainstream knowledge in the English-speaking literature. Studies in a non-Western context have potentials to generate new insights to the existing knowledge, but also can help to critically reflect on the mainstream Anglo-phone theories (Yeung, 2007). Therefore, my aims are not only simply to address the questions by providing some 'Chinese' empirical stories on the topic. But also, to shed some light on the very fundamental issue in (E)EG—namely the nature of regionally uneven economic evolution and long-term development—through a systematic comparison of several Chinese OICs from an integrated perspective in which both structure and agency factors and their broader nexus and relations embedded in and active at multiple geographical scales are considered. In doing so, three main research questions are posed as follows:

- ***Research Question 1: How is regional economic resilience conceptualized? What are the impact factors affecting regional economic resilience including the nature of various crises/shocks?***

The book aims to answer what kind of notions, concepts and paradigms in EG can be properly used to (re)frame a sound conceptual framework of regional economic resilience. I not only think the current account of regional economic resilience, also as the mainstream one, in the literature, is built upon a taken-for-granted prerequisite that is resilience is about regional endogenous structure-based ability and/or (multi-scalar) agency-led adaptability in the context of the 2008 financial crisis (or economic crises alike) (Hu et al., 2022; Tan et al., 2020). But also going beyond this EEG account, regional economic resilience is also relational and institutional, for which industrial, social and political relations and networks across space constitute the very nature of adaptation and adaptability for resilience evolution. And also, the nature of crisis, as Martin et al. (2016) pointed, give rise to new elements and perspectives to understand regional economic resilience more fundamentally, not merely focusing the traditional subject of resilience, namely local economies and their related actors and elements.

- ***Research Question 2: How do local economies evolve and adapt with crises, and to what extent are they able (or not) to change old industries and create new ones? And how and to what extent regional path development is linked to and give impact on regional economic resilience?***

Given the complexity of the 'how' question relating to the process/mechanism, as noted above, I will adopt a more integrated and multi-scalarity of research object perspective to explore the question. To do so, three key issues from different but interlinked scales will be closely focused: (1). at the micro level, the (re)actions of individuals from firms and organizations for the change or continuity in OICs over time;) (2) at the meso level, both industrial-sectorial and political-institutional agencies in forms of networks, institutions and policies that shape and reshape the adaptation and broader evolutionary paths of OICs; (3) at the macro level, national and international forces including investments, policy actions, market changes that impact on regional economic adaptability.

I am convinced that these three issues combined can help to understand better the nature of how OICs manage to response to crises and how they adapt for future development. In line with this multi-scalar view, and with an integrated conceptual perspective, the question is explored in Chaps. 6 and 7 with a focus on two Chinese coal-mining regions, in which the roles of multi-scalar politics, states and human agencies, among others, are emphasized, respectively.

- ***Research Question 3: why do Chinese OICs adapt differently to certain shocks on the ground and how to understand their unevenness in regional economic resilience in terms of process, mechanism and effectiveness on the ground?***

Precisely, why do some OICs seem more capable of renewing old industries and creating new growth paths while others are less able to or even fail to do that?

This question indeed has been one of the most intriguing issues, but also the most under-explored ones in economic geography (Martin & Sunley, 2006). On the basis of the work on the first question, this question aims to answer the meanings of adaptation and adaptability in regional economic resilience in a more systematic way, which more directly contribute to answer the 'why' question. The key aim of this book is to engage the notion of adaptation and adaptability, to adopt an EEG and relational perspective , into the critical question of uneven evolutionary geographies, trying to understand two three sub-questions: (1) is adaptation an actional process of adjustment on old existing industry path, to what extent and by what way it involves with the ability of developing new industry paths? that is, how to justify the relations between adaptation and adaptability to understand the nature of regional resilience in OICs? (2) If adaptability is about the ability to develop new growth paths for a sustainable perspective, how to identify the quality and effectiveness of this ability? and (3) in facing the crisis of COVID-19, what impact factors affect the differences in the economic resilience in Chinese OICs?

1.4 Book Structure and Overview of Chapters

This book is organized by eight chapters with a logical order, following from conceptual thinking to empirical exploration with a comparative setting (see Fig. 1.1). In the next chapter (Chap. 2), I mainly introduce the recent concerns and developments in the field of regional economic resilience. These concerns are about how regions adapt to a variety of challenges based on the notions of adaptation and adaptability and why their reflected resilience as a whole is geographically different with regard to the role of impact factors. Chapter 2 gives a conceptual overview of regional economic resilience, adaptation, and adaptability and suggests a more holistic conceptual framework.

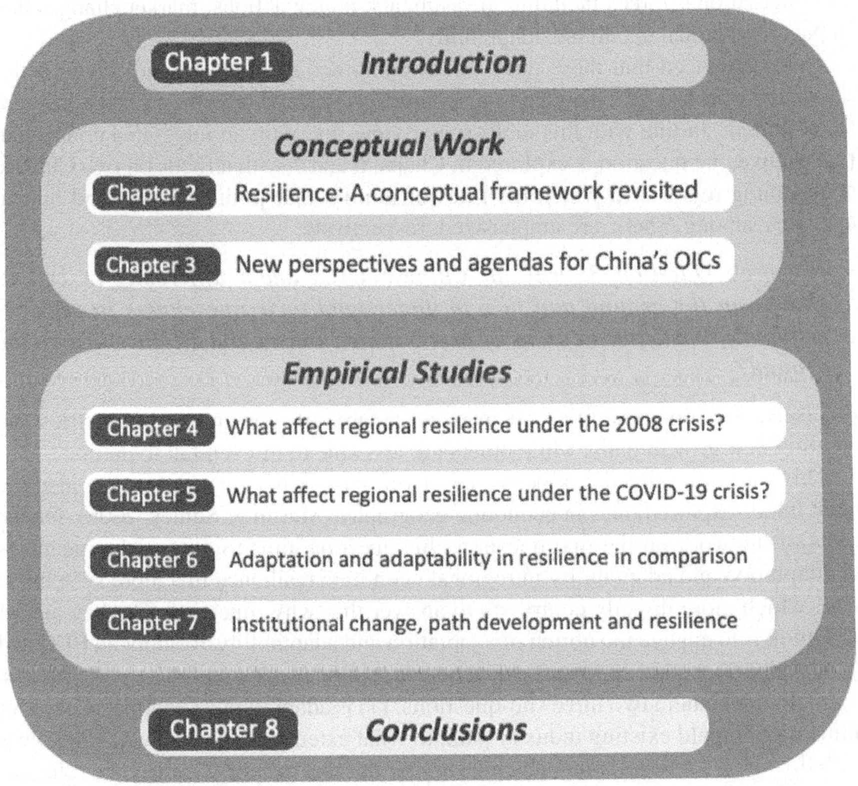

Fig. 1.1 The chapter structure of the book

Based on that, the book follows a collection of publications by the author, which consists of five journal articles (Chaps. 3–7), expect Chap. 8 as conclusions of the book. All chapters (including Chaps. 2–7) are academic papers with a double blinded peer-review process, which deal with both the conceptual and empirical work in the book (Fig. 1.2).

Chapter 3 engages with a future conceptual reflection on economic resilience from the notions of adaptation and adaptability in OICs by synthesizing a variety of concepts. These concepts are not only from EEG, but also from institutional and relational EG perspectives. Moreover, the chapter not only pays more attention to the traditional economy-related crises, but also give more emphasis on the recent, comprehensive, yet less examined crisis —the COVID-19 pandemic in the conceptualizations of regional economic resilience. This chapter thus clarities the potentials of adopting several internationally recognized concepts and perspectives in the Chinese-specific context, and also suggests some avenues for future research on the topic.

On the basis of the first four chapters, the empirical analyses are presented in Chaps. 4–7, which serve to answer two interlinked main research questions: first,

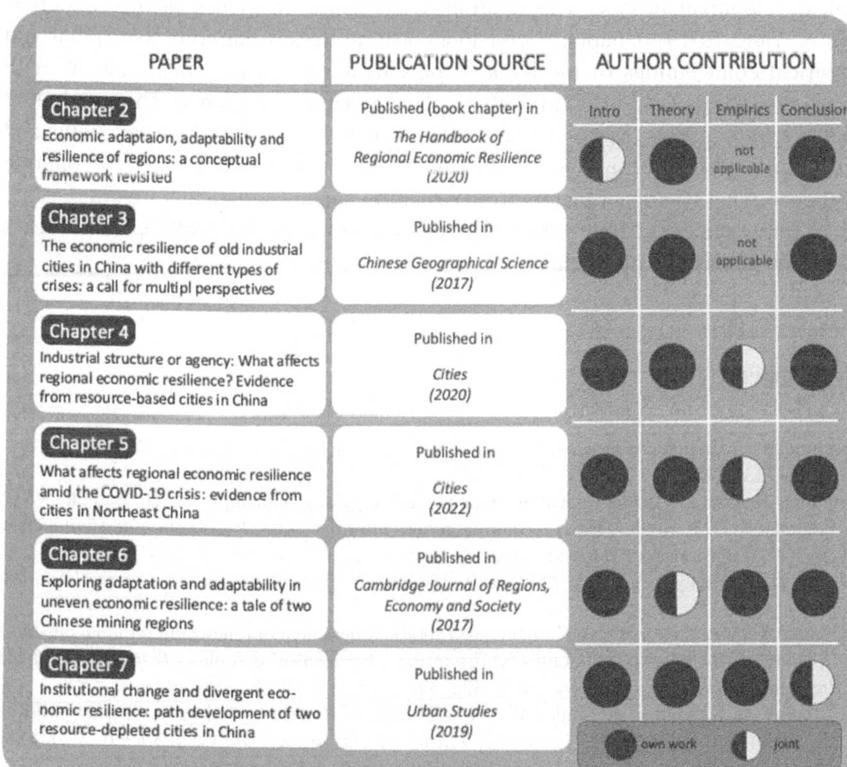

Fig. 1.2 Overview of published chapters and author's contribution

how do OICs in China adapt to a certain crisis on the ground, and why do different OICs adapt differently when facing the same crisis? Second, how to understand the differentiations in economic resilience for OICs when facing the crisis of the COVID-19 pandemic? What is the role of the nature of crisis in affecting regional economic resilience?

To be clear, Fig. 1.2 is drawn to give the details (including publication information, author contribution, and time of publication) to each individual paper. In short, Chaps. 4 and 5 delivers quantitative research on exploring the impact factors of regional economic resilience of OICs in Northeast China facing the 2008 financial crisis and the 2019 COVID-19 crisis, respectively (Hu et al., 2022; Tan et al., 2020). Chapters 6 and 7 provide comparative case studies on regional economic resilience—based on two Chinese typical coal-mining cities, namely Zaozhuang (Shandong) and Fuxin (Liaoning)—from different perspectives. Chapter 6 adopts an updated EEG perspective with the path development approach. It mainly demonstrates how varied forms of adaptation-adaptability relationships may result in different patterns of processes and also different potentials of resilience (Hu & Hassink, 2017). Chapter 7 Draws upon institutional change and path development concepts, and develops an analytical framework that explains how different modes of institutional change shape path development processes in relation to economic resilience in the two cities (Hu & Yang, 2019). Chapter 8 concludes and addresses the both conceptual and empirical contributions of the book. It puts forward a "engaged pluralism" with focus on EEG, REG and IEG in study regional economic resilience. The concluding chapter also suggests the co-evolving relationships between old industry paths and new industry paths are the keys to deepen the conceptualizations of regional economic resilience. That is, impact factors from structure, agency and the nature of crisis are all of importance to capture the real meaning and value of resilience in future's economic geography research.

References

Boschma, R. (2015). Towards an evolutionary perspective on regional resilience. *Regional Studies, 49*(5), 733–751.

Boschma, R. (2011). Technological relatedness and regional branching. In H. Bathelt, M. P. Feldman & D. F. Kogler (Eds.), *Dynamic geographies of knowledge creation and innovation* (pp. 64–81). Routlegde, Talor and Francis.

Bristow, G., & Healy, A. (2014). Regional resilience: An agency perspective. *Regional Studies, 48*(5), 923–935.

Chapman, K., MacKinnon, D., & Cumbers, A. (2004). Adjustment or renewal in regional clusters? A study of diversification amongst SMEs in the Aberdeen oil complex. *Transactions of the Institute of British Geographers*, 29(3), 382–396.

Christopherson, S., Martin, R., Sunley, P., & Tyler, P. (2014). Reindustrializing regions: Re-building the manufacturing economy? *Cambridge Journal of Regions, Economy and Society, 7*(3), 351–358.

Christopherson, S., Michie, J., & Tyler, P. (2010). Regional resilience: Theoretical and empirical perspectives. *Cambridge Journal of Regions, Economy and Society, 3*(1), 3–10.

Dawley, S. (2013). Creating new paths? Offshore wind, policy activism, and peripheral region development. *Economic Geography, 90*(1), 91–112.

Eich-Born, M., & Hassink, R. (2005). On the battle between shipbuilding regions in Germany and South Korea. *Environment and Planning A, 37*, 635–656.

Frenken, K., Van-Oort, F., & Verburg, T. (2007). Related variety, unrelated variety and regional economic growth. *Regional Studies, 41*(5), 685–697.

Garud, R., Kumaraswamy, A., & Karnøe, P. (2010). Path dependence or path creation? *Journal of Management Studies, 47*(4), 760–774.

Glaeser, E. L. (2022). Urban resilience. *Urban Studies, 59*(1), 3–35.

Grabher, G. (1993). The weakness of strong ties: The lock-in of regional development in the Ruhr Area. In G. Grabher (Ed.), *The embedded firm: On the socio-economics of industrial networks* (pp. 255–277). Routledge.

Grabher, G., & Stark, D. (1997). Organizing diversity: Evolutionary theory, network analysis and post-socialism. *Regional Studies, 31*(5), 533–544.

Hassink, R. (2007). The strength of weak lock-ins: the renewal of the Westmünsterland textile industry. *Environment and Planning A*, 39(5), 1147–1165.

Hassink, R. (2010a). Regional resilience: A promising concept to explain differences in regional economic adaptability? *Cambridge Journal of Regions, Economy and Society, 3*(1), 45–58.

Hassink, R. (2010b). Locked in decline? On the role of regional lock-ins in old industrial areas. In R. Boschma & R. Martin (Eds.), *The handbook of evolutionary economic geography* (pp. 450–468). Edward Elgar.

Hassink, R., Klaerding, C., & Marques, P. (2014). Advancing evolutionary economic geography by engaged pluralism. *Regional Studies, 48*(7), 1295–1307.

Hu, X. H. (2014). State-led path creation in China's rustbelt: The case of Fuxin. *Regional Studies, Regional Science, 1*(1), 294–300.

Hu, X., & Hassink, R. (2017). Exploring adaptation and adaptability in uneven economic resilience: A tale of two Chinese mining regions. *Cambridge Journal of Regions, Economy and Society, 10*(3), 527–541.

Hu, X., & Hassink, R. (2020). Adaptation, adaptability and regional economic resilience: A conceptual framework. In G. Bristow & A. Healy (Eds.), *Handbook on regional economic resilience* (pp. 54–68). Edward Elgar.

Hu, X., Li, L., & Dong, K. (2022). What matters for regional economic resilience amid COVID-19? Evidence from cities in Northeast China. *Cities, 120*, 103440.

Hu, X., & Yang, C. (2019). Institutional change and divergent economic resilience: Path development of two resource-depleted cities in China. *Urban Studies, 56*(16), 3466–3485.

Hudson, R. (2005). Rethinking change in old industrial regions: Reflecting on the experiences of North East England. *Environment and Planning A, 37*(4), 581–596.

Isaksen, A. (2014). Industrial development in thin regions: Trapped in path extension? *Journal of Economic Geography, 15*(3), 585–600.

MacKinnon, D., Cumbers, A., Pike, A., Birch, K., & McMaster, R. (2009). Evolution in economic geography: Institutions, political economy and adaptation. *Economic Geography, 85*(2), 129–150.

MacKinnon, D., & Derickson, K. D. (2013). From resilience to resourcefulness: A critique of resilience policy and activism. *Progress in Human Geography, 37*(2), 253–270.

Malecki, E. (2009). Geographical environments for entrepreneurship. *International Journal of Entrepreneurship and Small Business, 7*(2), 175–190.

Markusen, A. (2003). Fuzzy concepts, scanty evidence, policy distance: The case for rigour and policy relevance in critical regional studies. *Regional Studies, 37*(6–7), 701–717.

Martin, R. (2010). Rethinking regional path dependence: Beyond lock-in to evolution. *Economic Geography, 86*(1), 1–27.

Martin, R. (2012a). Regional economic resilience, hysteresis and recessionary shocks. *Journal of Economic Geography, 12*(1), 1–32.

Martin, R. (2012b). (Re)Placing path dependence: A response to the debate. *International Journal of Urban and Regional Research, 36*(1), 179–192.

Martin, R., & Simmie, J. (2008). Path dependence and local innovation systems in city-regions. *Management, Policy and Practice, 10*(2–3), 183–196.

Martin, R., & Sunley, P. (2006). Path dependence and regional economic evolution. *Journal of Economic Geography, 6*(4), 395–437.

Martin, R., & Sunley, P. (2015). On the notion of regional economic resilience: conceptualization and explanation. *Journal of Economic Geography*, 15(1), 1–42.

Martin, R., Sunley, P., Gardiner, B., & Tyler, P. (2016). How regions react to recessions: Resilience and the role of economic structure. *Regional Studies, 50*(4), 561–585.

Martin, R., & Sunley, P. (2020). *Regional economic resilience: Evolution and evaluation. Handbook on Regional Economic Resilience* 10–35.

Morgan, K. J. (2012). Path dependence and the state: The politics of novelty in old indus-trial regions. In P. Cooke (Ed.), *Re-framing Regional Development: Evolution, Innovation and Transition, Regions and Cities* (pp. 318–340). Routledge.

Neffke, F., Henning, M., & Boschma, R. (2011). How do regions diversify over time? Industry relatedness and the development of new growth paths in regions. *Economic Geography, 87*(3), 237–265

Nicola, M., Alsafi, Z., Sohrabi, C., et al. (2020). The socio-economic implications of the coronavirus pandemic (COVID-19): A review. *International Journal of Surgery, 78*, 185–193.

OECD. (2020). A systemic resilience approach to dealing with Covid-19 and future shocks. In *OECD policy responses to coronavirus (COVID-19)*. OECD Publishing.

Pike, A., Dawley, S., & Tomaney, J. (2010). Resilience, adaptation and adaptability. *Cambridge Journal of Regions, Economy and Society, 3*(1), 59–70.

Strambach, S., & Halkier, H. (2013). Reconceptualising change: Path dependency, path plasticity and knowledge combination. *Zeitschrift Fuer Wirtschaftsgeographie, 57*(1–2), 1–14.

Tan, J., Hu, X., Hassink, R., & Ni, J. (2020). Industrial structure or agency: What affects regional economic resilience? Evidence from resource-based cities in China. *Cities, 106*, 102906.

Yeung, H. W. C. (2007). Remaking economic geography: Insights from East Asia. *Economic Geography, 83*(4), 339–348.

Yeung, H. W. C., & Lin, G. C. S. (2003). Theorizing economic geographies of Asia. *Economic Geography, 79*(2), 107–128.

Chapter 2
Economic Adaptation, Adaptability, and Resilience of Regions: A Conceptual Framework Revisited

2.1 Introduction

In recent years, the notion of regional resilience has become a 'buzzword' in economic geography and regional studies (Martin & Sunley, 2015). On the one hand, its increasing popularity reflects the on-going 'recovery' policy agenda at multiple spatial levels after the 2008 global economic crisis (Foster, 2007). On the other hand, the notion's emphasis on history-related mechanisms of regional dynamics closely resonates with the 'evolutionary turn' in economic geography (Boschma, 2015; Hassink, 2010a). By rejecting the conventional (multi-) equilibrium view, the notion basically offers a promising intellectual tool to analyze and understand how a regional economy responds to and recovers from shocks or disturbances in changing dynamic contexts (Martin, 2012). However, the understanding of the notion is not unproblematic. Many studies consider resilience as a metaphor, namely a reductionist analogy between the workings of nature and those of man-made regional economic systems, rather than a thorough concept (Pendall et al., 2010). Moreover, in many cases, the resilience term is used in a loose way, either as a fashionable policy instrument, or as a wishful thought for future development (Martin & Sunley, 2015). This has raised worries that regional resilience might become a 'fuzzy concept' akin to 'clusters' (Christopherson et al., 2010: p. 4; Gong & Hassink, 2017), losing its conceptual value and durability in future research.

Leading economic geographers believe that regional resilience should not be a transient notion, despite its variegated conceptual roots (Martin, 2012; Martin & Sunley, 2015). Christopherson et al., (2010, p. 3) point out that resilience echoes, in a fundamental sense, "a very old and enduring question" in economic geography, that is, why some regional economies are able to cope with all sorts of shocks

(This chapter is a reprint version with slight differences from Hu, X., & Hassink, R. (2020). Adaptation, adaptability and regional economic resilience: A conceptual framework. Handbook on Regional Economic Resilience, 54–68.)

X. Hu, *Uneven Economic Resilience of Old Industrial Cities in China*,
Urban Sustainability, https://doi.org/10.1007/978-981-99-9279-9_2

towards sustainable growth while others are not but locked in decline (Hassink, 2010a; Simmie & Martin, 2010). Furthermore, as suggested by Martin and Sunley (2015, p. 3), "regional resilience is part of the process of geographically uneven economic evolution and development, and should be theorized and analyzed as such".

Informed by the complex adaptive systems theory, adaptation and adaptability are the most debated notions in urban and regional resilience research (Boschma, 2015; Hassink, 2010a; Martin, 2012; Oliva & Lazzeretti, 2017; Pike et al., 2010). In many cases, adaptation is portrayed as a path dependent process maintaining existing development paths or primary functionalities of a system (Adger, 2010). It often involves short-term quantitative, non-radical changes or adjustments of the regional agenda, strategy and action, in order to regain or maintain the economy's 'core performances' (Martin et al., 2016), while adaptability is believed to be a more 'transformation-over-stabilization' term. The latter often refers to a long-term adaptive ability of an economic system seeking qualitative changes or shifting of industrial paths, developmental modes and systematic institutions (Boschma, 2015; Chapman et al., 2004; Pike et al., 2010). These assumptions, despite good reasons to support (Adger, 2010; Grabher, 1995; Grabher & Stark, 1997), lack rigor, which might undermine the conceptualization and operationalization of regional resilience.

Adaptation and adaptability are somehow preconditioned with competing meanings, but they are by no means separated. A number of studies indicate that the tensions between the two are dynamic, and that should be at the core when exploring the geographies of resilience. Boschma (2015, p. 734) argues that resilience is about "how to secure adaptation and adaptability simultaneously", and related variety is the key to strike such an industrial or institutional balance to make a region resilient in a long-run context. Moreover, according to Ibert and Schmidt (2014), adaptation and adaptability are integral dimensions of resilience. They can be coherently connected, when, for instance, strategic actors enact 'purposive adaptation' that nurtures adaptability for regional industrial path shifting and upgrading (Tomlinson & Branston, 2014). As has been emphasized by Pike et al. (2010), MacKinnon and Derickson (2013) in a Western context, the relationship between adaptation and adaptability is complex and dynamic, which is not only shaped by firm agency and embedded in a regional context, but also mediated by the interplay between the state, capital, labor and society in a broader multi-scalar institutional environment. The under-explored question here is not about whether adaptation and adaptability are related, but about how they interact, and how possibly different forms and extents of interaction could lead to differences in regional economic resilience.

Therefore, the purpose of this chapter, which draws from Hu and Hassink (2017a), is to advance the understanding of regional economic resilience. To avoid raising the confusion about the notion, this chapter incorporates a conceptualization into existing accounts on adaptation and adaptability, instead of introducing alternative terms or notions. We argue that current definitions of adaptation and adaptability have been suffering from the conventional wisdom of path dependence and lock-in (Grabher, 1993; Grabher & Stark, 1997), which need to be critically revisited. In doing so, the chapter (re)defines them in a related and interactive perspective. We suggest a typology of the adaptation-adaptability relationship. Following this, a conceptual framework is proposed in order to explain and analyze the unevenness of regional economic resilience. In this framework, we provide existing empirical examples as illustrations to show how changing multi-scalar factors and determinants shape out varieties of adaptation-adaptability relationships that matter for the regional differentiation of economic resilience.

Our chapter is organized as follows: Sect. 2.2 critically reviews the existing literature on regional economic resilience. It particularly focuses on the conceptual fuzziness and the inappropriate trade-off idea of adaptation and adaptability. Based on this, Sect. 2.3 puts forward a typology of the relationships between adaptation and adaptability. A subsequent set of impact factors and determinants affecting regional economic resilience is discussed in Sect. 2.4. Finally, Sect. 2.5 summarizes the findings and highlights our key arguments, suggesting that adaptation and adaptability should not be understood separately in any case. It is further argued that they are equally important to long-term regional economic evolution, and should be conceptualized as dialectical, dynamic, indicative dimensions of regional economic resilience.

2.2 Regional Economic Resilience: A Critical Review

The notion of *regional economic resilience* can be seen as a result of the analogy between the workings of nature and those of man-made regional economic systems in a post-crisis context. Akin to ecological and engineering resilience, regional economic resilience, initially, was associated with a local economic resistibility to, and also recoverability, from (sudden) shocks (Foster, 2007; Hill et al., 2008). This understanding, however, has been increasingly criticized. In a special issue on "The Regional Resilience" (Cambridge Journal of Regions, Economy, and Society, 3, 2010), it is suggested that regional economic resilience needs to reject the (multi-) equilibrium thinking adopted by ecology, psychology and disaster management, and to embrace evolutionary and systematic complex theories. Thus far, most studies regard resilience more as a metaphor, rather than a well-developed concept (Pendall et al., 2010). There is also a tendency to use it as a regional policy instrument or a development agenda. Despite a few insightful conceptualizations, the notion has suffered from fuzzy definitions and incoherent understandings.

In our reading, there are two main intertwined dimensions that have put the resilience notion in danger of becoming a 'fuzzy concept' (Markusen, 1999), namely, (1) varieties of definitions and shifting understandings and (2) the use of trade-off terminologies in conceptualization. Both are critically discussed as follows so as to identify the problems and shortcomings of the existing conceptualizations on resilience, and thus to point out potential avenues for future theoretical and empirical improvement of research .

(1) Varieties of definitions and shifting understandings

Resilience is a contextualized idea, suggesting the centrality of the 'recovery' or 'survival' discourse. It has been generally accepted as an ability of a system responding to a shock, bearing a clear task of maintaining stability and durability of the system. From this point, the 'bounce back' thinking seems reasonable when applying it to the regional economic system context. Because, recovery is the first priority of all tasks in a region after a shock and is also the most realistic agenda. Therefore, the original idea of resilience, by default, is based on particular conditions, restricted to a very-moment, recovery-oriented, and equilibrium-like setting.

Many economic geographers have tried to decontextualize regional economic resilience and put it into a broader perspective. It is defined not only as an ability to cope with one-off shocks or disturbances, but also as a long-term ability to adapt to omnipresent uncertainties (Martin, 2012; Martin & Sunley, 2015; Simmie & Martin, 2010). One reason is that some crises are not restricted to sudden shocks, but can be regarded as chronic problems and long-term challenges, such as deindustrialization, emigration, resource depletion and aging (Dawley et al., 2010). In a deeper sense, shocks have permanent or at least long-lasting effects in nature, and there is 'memory' of the shock imprinted itself in the economy, known as 'remanence' (Cross et al, 2009). In this regard, crises can be seen as an input of new conditions, making the existing context dynamic but still evolving in a path-dependent way. Regional economic resilience therefore matters for how a regional economy adapts to changing contexts other than merely to one or two specific shocks/crises. This perspective shows that resilience also makes sense for long-term processes of regional development (Martin & Sunley, 2015).

By drawing insights from complex adaptive systems theory, regional economic resilience has been further conceptualized as a dynamic adaptive ability. One strand of this direction centers on the four-phase adaptive cycle in which the degree of resilience varies according to the 'connectedness' of actors in a regional system (Pendall et al., 2010; Simmie & Martin, 2010). It is claimed that regional resilience evolves by following a circular order from 'conservation' and 'release' to 'reorganization' and 'exploitation', and back. This conceptualization is, however, rigid, not only because it tends to deliver a multiple equilibrium expression, but also because the factors shaping resilience are in reality not bounded to local endogenous networks. As an improvement, Martin (2012) combines the idea of hysteresis with resilience, suggesting four dimensions, namely 'resistance', 'recovery', 'reorientation' and 'renewal'. He highlights the varieties of possibilities about the reactions of regions to recessionary shocks, and the important roles of both shock characteristics and economic structural factors in the processes. Although, as Martin (2012) stressed, these dimensions are inter-related, it is, for example, under-examined how resistance is linked to renewal and to what extent recovery can be turned into a source for reorientation. To address this question, Martin and Sunley (2015) stress that resilience is a continuous process that consists of four main inter-connected types of abilities: 'vulnerability', 'resistance', 'robustness' and 'recoverability'. The notion of robustness, for example, conditioned by the former ones, is concerned with regaining key functions or performance of a system, but at the same time, flexibly allowing the changes of structural elements and ways of doing things for long-term regional competitiveness (Martin & Sunley, 2015, 2017). However, the term 'recoverability', again, in our view, reminds us of resilience for a 'short-term' recovery agenda rather than long-term evolution.

All in all, the definition of resilience seems highly adaptable. It varies with the advancement of understandings of regional economies as continuously evolving, open-ended and complex systems. There is no coherently agreed core meaning of regional economic resilience. Although some have tended to detach regional resilience from the 'recovery' context, others would ask: what is the aim of resilience and how to operationalize it in everyday economic life? This question concerning the need to (de)contextualize resilience is still under-explored. Moreover, the existing interpretations are heavily influenced by a variety of disciplines, resulting in a trend of an increasing adoption of new terminologies, while lacking clear explanations (MacKinnon & Derickson, 2013). Thus, in our view, the future conceptualization should aim to integrate new perspectives or approaches into existing evolutionary concepts by asking fundamental questions, instead of simply and uncritically borrowing new ideas from other disciplines. As warned by Hassink (2010a, p. 55), "we should be careful to take up another fashionable concept without first carefully scrutinizing the value of it in answering key questions of our subject."

(2) The use of trade-off thinking

There have been a number of regional resilience studies in favor of adopting incompatible dual terms, such as 'bounce back' or 'move on', 'lock-in' or 'lock out', 'recovery' or 're-orientation' (Cowell, 2013; Crespo et al., 2014; Martin, 2012). Most of them are utilized in a shallow way, lacking thorough concerns on the interrelations between the terms. This has been somehow common, not merely in the resilience research, but also in broader urban and regional studies (Sayer, 1991). That is, when a specific term or notion is used, the opposite or incompatible one is likely to be coupled with it. Without specific explanation, such dualism often gives us the impression that a regional economy as a homogenous entity evolves either in this way or in that way, as if there is nothing in between. Moreover, it seems that once a trade-off idea is introduced, it has a long-lasting effect that is difficult to be removed. This can be exemplified by one of the most often used dualistic terms in the regional resilience literature, namely adaptation and adaptability.

Originally stemming from Grabher (1993), Grabher and Stark (1997) conceptualized adaptation and adaptability for regional firm dynamics and emphasized that there is a strong trade-off idea between them. Drawing from the idea of 'the weakness of strong ties' from economic sociology, they argued that strongly tied social agents with a cohesive action can facilitate adaptation in the short run, but hinder adaptability in the long run. Because adaptation involves "exploiting known territory at the cost of forgetting (or never learning) the skills of exploring for new solutions" (Grabher & Stark, 1997, p. 534). A similar point can be captured in Martin and Sunley (2006, p. 421)'s argument: "there is a trade-off between specialization and a short-lived burst of fast regional growth on the one hand, and diversity and continual regional adaptability on the other". Adaptation does play a certain constraining role in adaptability and vice versa. Such a trade-off idea, however, is creating a clear-cut dualism in which adaptation is in opposition to adaptability. In many cases, adaptation refers to short-term incremental changes within the old paths, while adaptability is more related to radical change, innovation and new path creation (Chapman et al., 2004; Kirsh, 1996).

When using adaptation and adaptability to conceptualize regional resilience, the dualism thinking has become a key cause for conceptual superficiality. In the literature, adaptation tends to be automatically linked to a low-level resilience, echoing with the term of 'resistance' and 'recovery'. While, adaptability often means high-level resilience related to 'reorientation' and 'renewal'. Even more, many argue that regional resilience is about a long-term capacity to foster structural change, "rather than to continue doing the same thing" (Adger, 2010, p. 1; Folke, 2006). They therefore prefer using adaptability to indicate resilience, and reject adaptation for its predestinated pathway towards 'equilibrium' or 'lock-in'. Others, however, and in particular evolutionary economic geographers, view resilience as a dynamic process. They argue, that, under some conditions, adaptation can turn out positive feedbacks for adaptability, leading to regional renewal and diversification (Carlsson et al., 2014; Martin, 2012; Martin & Sunley, 2006). Martin and Sunley (2015) further think that adaptation relating to coping with short-run shocks should be regarded as important as adaptability for long-term economic evolution. Given this, it is suggested that the future conceptualization on regional resilience should center on understanding the interactions and relationships between adaptation and adaptability (Boschma, 2015; Pike et al., 2010).

However, the dualistic thinking still plays a dominant role in understanding regional resilience. Such thinking is basically derived from a neglect of the dynamic role of history. Adaptation and adaptability were initially defined to serve for the rigid and static argument of 'once locked-in, never get out' (Grabher, 1993). This has led to a divergence that on the adaptation side, regional economic resilience means to stick to path dependence, while, on the adaptability side, to completely detach from path dependence. This means that pursuing adaptation is at the cost of fostering adaptability. Such understanding is highly problematic, as Sayer (1991: p. 287) states, "the [trade-off] terms not only oppose but presuppose one another in some respects ... certain of their senses may be internally rather than externally related...[when] they are internally related they can justifiably be said to be dialectical". This gives a hint that adaptation and adaptability are more or less inter-related in nature and that they should be defined in a logically related and interactive setting. Thus, to deconstruct the adaptation-adaptability trade-off and to justify them in a dialectical way is important and highly needed.

2.3 Placing Adaptation and Adaptability in Regional Economic Resilience: A Typology

The dominant definition of adaptation is strongly preconditioned by a short-term development context. This however gives a false expression, as if adaptation did not persist for a long-term period, and as if it was merely about a responsive action or way of action. In our view, adaptation should be regarded as an integral part of long-term regional economic evolution. First, adaptation occurs in a continuous rather than a punctuated way. Actors adapt not only to specific short-term shocks, but also to ubiquitous uncertainties in a changing context. Secondly, history is an important input for adaptation, but never plays a determining role. It can be regarded as an inherited characteristic seeking to maintain or restore the previous performance or functionality of a socio-economic system. But the ways in which adaptation unfolds are open-ended, which can lead to divergent developmental processes and pathways within a region.

Compared to adaptation, adaptability is often understood as the major indicator of, sometimes even equivalent to, resilience. A deficiency in this understanding is that adaptability is separated from adaptation and always positively affects resilience. In our view, adaptability does not emerge from scratch. It rather emerges out of an adaptive evolutionary process. In other words, adaptability is not given, but is, to a greater or lesser extent, built upon the practices of adaptation. Adaptation and adaptability can, and more likely co-exist and co-evolve in a dynamic way, and should be conceptualized as such (MacKinnon et al., 2009).

We would therefore suggest an updated definition of adaptation and adaptability:

Adaptation is an on-going and never-ending process, by which a regional economic system response to a succession of challenges and disturbances, in order to fit to its varying environment. It refers to both the current function/performance of being maintained and to the potentials for regional dynamics.

Adaptability is the ability to create new and/or change old actors, institutions and resources in a regional economy, which involves an action of innovation. It is a result of adaptation in which some actors intentionally or unintentionally adapt to environments. In a long-term perspective, adaptability should aim to strive for alternative modes of doing things, rather than maintaining the existing functions and structures.

Based on these definitions, we would stress that the long-term evolution of a regional economy will most likely involve with both adaptation and adaptability. And the ways in which they interact over time are indicative of the differentiation of regional economic resilience. Further, we identify at least five possible types of relationships between adaptation and adaptability and address how different types of relationships shape regional economic resilience differently. These types are: (1) they are against each other, (2) they keep apart from each other, (3) adaptation enables adaptability, (4) adaptation benefits from adaptability, and (5) they are mutually complementary.

Type (1): They are against each other

Adaptability may not necessarily lead to a structural change of the local economy. It is possible that regions achieve adaptability at the cost of adaptation, and such adaptability itself is compromised by adaptation. This situation may be more likely to occur in regions where old and new industries co-exist in a cutthroat competition relationship. These new industries may be quickly fostered by actors in response to external opportunities such as policy interventions or market dynamics, which are detached from local existing economic structures (Hu, 2017). They may face problems once the external environment changes and tend to survive by occupying limited local assets. This is likely to lead to a lose-lose situation of adaptation and adaptability negatively affecting each other.

Many old industrial areas, for instance, are keen to introduce exogenous knowledge, technologies and labor for path creation as compensation to the loss of old industries (Dawley, 2014; Isaksen, 2015). But it is difficult for exogenous sources to permeate local established industrial niches, given the lack of social learning and interaction (Coenen et al., 2015). Regional economies can also become 'locked in' to global production networks controlled by powerful transnational corporations (MacKinnon, 2011). These regions' adaptation may become internally divergent, in which actors enact with incoherent interests and powers (Dawley et al., 2010). The competing relationship between adaptation and adaptability may also exist in regions where local policy-makers frequently shift development strategies/policies to fit to changing contexts, but with limited concerns about existing industries (Peck & Theodore, 2015). Although in this context adaptability may involve new path creation, it is however shortsighted, being difficult to trigger substantial change. This relationship is subject to harmful resilience, which would undermine long-term regional evolution.

Type (2): They keep apart from each other

Some regions' adaptation may strongly involve introducing external technologies, capital and industries, rather than significant changes within existing industries or paths. This might be perhaps caused by the institutional difficulties or resistance to modernize old industries, also perhaps by the eagerness to diversify regional economies in order to compensate for the loss of declining industries. Within this context, adaptability may have a strong purpose: to introduce new industries from elsewhere that are not at the cost of local core industries but can be beneficial to the overall economy. One typical example is that regions achieve 'unrelated variety' by mainly attracting new industries that are not technologically related to the existing ones (Frenken et al., 2007). These industries might be vertically integrated and thus horizontally separated from local economic systems. They basically run in parallel with the existing industries. Such adaptability may result in a highly incoherent regional evolution, resulting in multiple heterogeneous adaptive trajectories (Boschma, 2015). As such, adaptability might be enhanced by an increase in industrial/institutional heterogeneity, but adaptation (the core institutions and their main functions) stays virtually unchanged. The separation may result in a highly incoherent institutional milieu, preventing actors from aligning and combining resources for building up industrial specializations. A lack of industrial focus would potentially lead to undermine regional advantages and weaken long-term resilience.

Type (3): Adaptation enables adaptability

In comparison to the two relationships above, this type of relationship entails that local actors involved in adaptation consciously mobilize resources to facilitate adaptability. In general, two main mechanisms of practice can be distinguished. That is, adaptation can enable adaptability in an endogenous and exogenous way. The first mechanism involves constructive mobility from existing industrial assets to new uses. New industries or paths can be, for instance, primarily built upon place-specific resources, legacies and networks. In other words, they can branch out of local existing industrial domains. Such adaptability has a strong potential towards regional renewal (Neffke et al., 2011). In the second mechanism, however, adaptability may prove to depend on the ability of local actors to exploit and introduce extra-local forces for local economic dynamics. It may involve an action of attracting inward investment for building up new industries or a process of turning external opportunities into local endogenous impetus for industrial dynamics (Hu & Hassink, 2017b). We do not deny that in some circumstances adaptability supported by adaptation can generate positive feedbacks to reshape the previous form of adaptation. But we want to emphasize that it is possible that adaptation merely enables adaptability, particularly when local actors purposively foster expected new industries and lack efforts to restructure existing patterns and functionalities of economic development.

Type (4): Adaptation benefits from adaptability

Within this relationship, adaptation benefits from adaptability, but adaptation has limited impact on adaptability. Adaptability may be initially triggered by one-off contingency, which is little related to purposive adaptation. For example, Schumpeterian radical innovation may occur, albeit not on initial purpose. It may generate unexpected effects, leading to a path-breaking or de-locking process that fundamentally changes the whole ways of doing things (Martin & Sunley, 2006). This means, existing adaptation itself can be reshaped by such kind of adaptability and regional resilience thus would become stronger than before. However, adaptability, in some 'emergent' circumstances, may merely serve to sustain adaptation (Hu, 2014). Adaptability might be a part of the adaptation story, for example, if it is aimed to fix old problems, or if it is designed by non-local actors to protect inherited institutional architectures. In this sense, adaptability may not be necessarily beneficial to improve regional economic resilience.

Type (5): They are mutually complementary

This type represents an interactive and intertwined relationship between adaptation and adaptability. In some conditions adaptation and adaptability can be harmonized in a reciprocal manner, which is conducive to long-term regional resilience (Ibert & Schmidt, 2014; Tomlinson & Branston, 2014). Science-based innovation, spin-offs, and entrepreneurship can be seen as one example that can lead to positive adaptability and subsequently can reshape the previous form of regional adaptation (Feldman, 2007). In another case, some regional actors are aware of how to exploit global opportunities by strategically making existing regional assets 'plugged' into global production networks, leading to regional value creation and industrial upgrading (MacKinnon, 2011). This is a typical example to manifest that adaptation can enable adaptability through the role of strategic agency, and adaptability can in turn feedback to adaptation resulting in path renewal, for instance.

Additionally, some regions, particularly 'institutionally thin' peripheral regions, are not necessarily related to lock-ins and weak resilience. By analyzing the formation of the Örnsköldsvik bio-refinery industry in Sweden, Arbuthnott et al. (2010), for instance, highlight the importance of social interaction processes between traditional and new industry actors for reframing complementary business concepts in the region that helped to engender industrial renewal. Adaptability, albeit exogenously sourced, can be nurtured locally, in order to be more place-embedded, generating long-lasting positive impacts. It is evidenced that the state can act as a mediator actively involving the recombination of local and non-local knowledge, through policy interventions (Dawley, 2014; Isaksen & Trippl, 2016). As such, adaptability benefits from the existing adaptation process and simultaneously pumps transformative stimuli into the existing process of adaptation. Such type involves a co-evolutionary and dialectical process towards institutionally coherent and industrially complementary regional restructuring. This type can potentially form strong resilience.

To sum up, these different patterns of adaptation-adaptability relationships can shed light on differentiated economic resilience among regions. In reality, the types of relationships might be more complex than we proposed above. They differ over time and space, which are shaped by multi-actors and multi-scalar impact factors in broader social, economic and institutional environments (Giannakis & Bruggeman, 2017; Hassink, 2010b).

2.4 Impact Factors and Determinants

The varieties of adaptation-adaptability relationships provide a useful perspective to analyze regional resilience. But, to better understand it, we need to get a grasp of the factors affecting the adaptation-adaptability relationships and how they mold them differently among regions. Scholars have sought to identify the main determinants of adaptation and adaptability, albeit from different perspectives. Evolutionary economic geographers argue that endogenous factors such as the connectedness of social agents, local firm characteristics, industrial diversity and institutional structures are the core factors (Boschma, 2015; Brown & Greenbaum, 2017; Simmie & Martin, 2010). Others, however, state that uneven resilience is shaped by the position and relationship of regions and their actors within multi-scalar institutional environments, policy arrangements and changing political economy frameworks (Chapman et al, 2004; MacKinnon & Derickson, 2013; Pike et al., 2010). In our view, resilience is shaped by mixed impact factors, including economic-structural and political-institutional factors, characterized by endogenous and exogenous dynamics. These factors are path and place dependent, but also history-sensitive and context-emergent (Martin & Sunley, 2015; Nyström, 2017).

Two major impact factors can be identified, namely structural impact factors and contextual impact factors (Fig. 2.1). First, structural factors provide the bounded scope and set the constraints for the operation of embedded actors. For example, at the local level, the economic structure, the labor profile, governance framework and political ecology render local agents rational, pressuring them to practice a path-dependent adaptation process. Besides, local decision-making and actions are also affected by trans-local structural factors. A region's position in the global division of labor, for instance, may hinder the economy to upgrade or reorient into an advanced one in the short run, despite strong local advantages. Second, contextual factors refer to a current of dynamics, ranging from unpredictable shocks and events, to systemically (re)produced consequences. The emergence of new elements and stimuli can make the previous selection environment dynamic. Agents may be able to identify new opportunities and extend their scope and vision accordingly to nurture adaptability.

However, the two major impact factors are not in isolation. Rather, and indeed, they are not only horizontally entwined, but also are vertically related across space in a logical way (Fig. 2.1). These factors do not necessarily give a direct impact

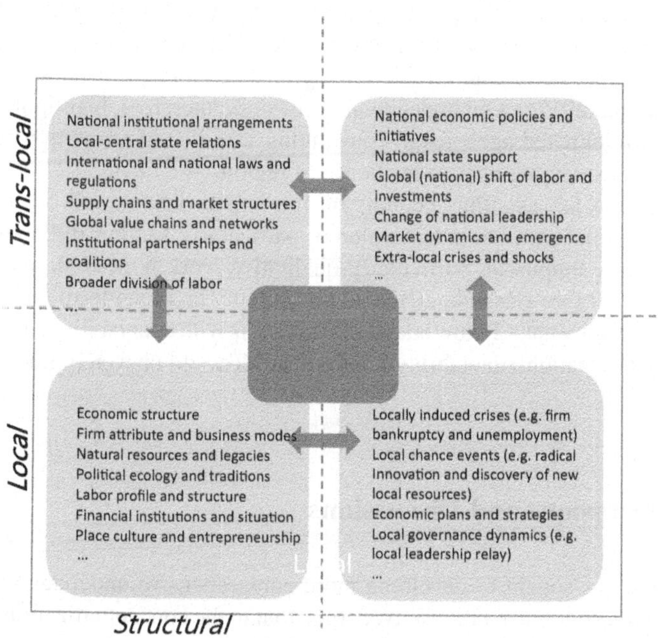

Fig. 2.1 Impact factors of regional resilience. *Source* adapted from Martin and Sunley (2015)

on adaptation and adaptability, but always, through influencing the actions of local agents and communities (Davidson, 2010; Robinson & Carson, 2015). Regions differ in resilience not merely because of the spatial differentiation of impact factors themselves, but largely depending on how local agents read, translate and value varied impact factors, and on how they act in practice accordingly (Bristow & Healy, 2014). In other words, the ways in which adaptation relates to adaptability are strongly linked to how local actors comprehensively scan a complex multi-scalar environment and to the adaptive action they carry out. Therefore, we argue that regional economic resilience matters for local agency, particularly in terms of the pattern and efficiency of responding to place-specific particular factors.

Of all impact factors of regional resilience, determinants are the most sensitive ones that can potentially answer the question of "adaptation to what" or "adaptation for what" (Smit et al., 2000). They play a key role in making a region's adaptation-adaptability relationship specific. They are, from an agency perspective, strongly linked to the interests of individuals and groups (such as entrepreneurs, government officials and stakeholders) (Bristow & Healy, 2014). Determinants differ in space, because local actors weigh the importance of impact factors differently. A national policy initiative, for example, may lead to different regional reactions: Some regions may take it as an opportunity and turn it into an impetus for industrial renewal, perhaps because local actors view the initiative as a good match with local specific conditions. Others may be less enthusiastic about it, and rather stick to old growth paths, because local actors might be more interested in or incentivized by other factors. Of course, we are not arguing that local actors and their agency always play a dominate role in shaping and reshaping regional economic resilience. Extra-local agency can make significant efforts to a regional economy, if local agency is weak (Hu & Hassink, 2017c). Our conceptual framework, therefore, highlights the role of spatially differentiated agencies in constituting varieties of adaptation-adaptability relationships explaining uneven resilience of regions.

However, so far, existing accounts on the impact factors and determinants of regional resilience remain under-explored, particularly lacking in-depth empirical evidence. More empirical, particularly qualitative work is needed, also in a non-Western context (see Hassink, 2017), to understand, in a long-term perspective, (1) who adapts to/for what, (2) how adaptation interacts with adaptability and (3) why the diversity in adaptation-adaptability relationships leads to uneven regional economic resilience.

2.5 Conclusions and Discussions

This chapter has sought to develop a more comprehensive understanding of long-term regional economic resilience. We argue that in the conceptualization of regional resilience there is a need to avoid dualism thinking and adopt an ontologically coherent evolutionary perspective. In doing so, a new conceptual framework for explaining uneven regional resilience is developed. In this framework, we reject the trade-off idea of adaptation and adaptability, and redefine both terms in a dialectical and interactive way.

It is argued that regional economic resilience should not be restricted to short-term recoverability or the ability related to resistance during or after shocks. It can and should be regarded as a long-term process of adaptation in an uncertain and post-crisis context. Such adaptation consists of varieties of heterogeneous actions and interest within a regional economy. Adaptability is a result, and also an integral part of the adaptation process. It can refer to path creation for temporary purpose serving to old structures. It also can refer to the ability to develop long-term regional economic sustainability and transformability.

A further conceptualization has been made in this chapter, suggesting that there are varieties of adaptation-adaptability relationships. These relationships are indicative of the geographically uneven processes, patterns and degrees of resilience. Following these arguments, we turned to an in-depth discussion of what impact factors affect and mold certain specific forms of adaptation-adaptability relationships in regions. It is stressed that adaptation-adaptability relationships vary from region to region, and are shaped by both contextual and structural impact factors at multi-scalar spatial levels.

We suggest that the definition of adaptation in relation to changes within pre-existing paths and adaptability involving new path creation needs to be revisited. In a long-term perspective, new paths may be created as a strategic response to cope with old problems. These new paths might be disconnected to existing paths and might sustain for a long-term period of time. Therefore, new path creations should not always be regarded as positive to regional evolution. Whether being positive or not should be contingent on how they are related to old existing structures of adaptation and to what extent they are embedded to the whole industrial system. In our view, high-quality regional resilience refers to a complementary, symbiotic, place-based relationship between adaptation and adaptability. Regional resilience is not merely about the ability and performance of overcoming shocks. It is rather about long-term endogenous (or endogenously and exogenously well balanced) quality, transformability and sustainability of regional evolution and development.

To end up, this research generates several policy advices regarding to regional economic resilience. Regional economic resilience should not be simply used as policy agenda without closely looking at place-based specificities of industrial, social and political settings and structures. Particularly, policy-makers should avoid a quick policy fix by, for instance, instrumentally adding new industries onto the old industrial systems (Coenen et al., 2015). An evolutionary perspective could be useful for national policy-makers to refine their existing preferential policies on mature regional economies. Future policies should focus more on how to intentionally cultivate the endogenous ability of local firms and industries to develop place-specific strategies, rather than to directly transplant new 'planned' industries/paths. For instance, policies do not necessarily have to develop a completely new economic make-up in post-crisis regions, but may be able to draw on existing assets, institutions and competencies (Bathelt et al., 2013; Boschma & Lambooy, 1999). If national policies try to promote healthy and robust industrial diversification in regions, fine-tuning policies are critically needed, to ensure that new paths are simply related to, but also beneficial to the dynamics of prevailing paths and structures.

Acknowledgements Robert Hassink from Kiel University has made great contributions to this chapter. We express our heartfelt thanks to him.

References

Adger, N. (2010). An interview with Neil Adger: resilience, adaptability, localization and transition. Transition culture. Retrieved December, 12 2016, at transitionculuture.org/2010/ 03/26/ an-interview-with-neil-adger-resilience-adaptability-localization-and-transition/

Arbuthnott, A., Eriksson, J., & Wincent, J. (2010). When a new industry meets traditional and declining ones: An integrative approach towards dialectics and social movement theory in a model of regional industry emergence processes. *Scandinavian Journal of Management, 26*(3), 290–308.

Bathelt, H., Munro, A. K., & Spigel, B. (2013). Challenges of transformation: Innovation, re-bundling and traditional manufacturing in Canada's technology triangle. *Regional Studies, 47*(7), 1111–1130.

Boschma, R. (2015). Towards an evolutionary perspective on regional resilience. *Regional Studies, 49*(5), 733–751.

Boschma, R., & Lambooy, J. (1999). The prospects of an adjustment policy based on collective learning in old industrial regions. *GeoJournal, 49*(4), 391–399.

Bristow, G., & Healy, A. (2014). Regional resilience: An agency perspective. *Regional Studies, 48*(5), 923–935.

Brown, L., & Greenbaum, R. T. (2017). The role of industrial diversity in economic resilience: An empirical examination across 35 years. *Urban Studies, 54*(6), 1347–1366.

Carlsson, E., Steen, M., Sand, R., & Nilsen, S. K. (2014). Resilient peripheral regions? The long-term effects of ten Norwegian restructuring programmes. *Norsk Geografisk Tidsskrift-Norwegian Journal of Geography, 68*(2), 91–101.

Chapman, K., MacKinnon, D., & Cumbers, A. (2004). Adjustment or renewal in regional clusters? A study of diversification amongst SMEs in the Aberdeen oil complex. *Transactions of the Institute of British Geographers, 29*(3), 382–396.

Christopherson, S., Michie, J., & Tyler, P. (2010). Regional resilience: Theoretical and empirical perspectives. *Cambridge Journal of Regions, Economy and Society, 3*(1), 3–10.

Coenen, L., Moodysson, J., & Martin, H. (2015). Path renewal in old industrial regions: Possibilities and limitations for regional innovation policy. *Regional Studies, 49*(5), 850–865.

Cowell, M. M. (2013). Bounce back or move on: Regional resilience and economic development planning. *Cities, 30*(3), 212–222.

Crespo, J., Suire, R., & Vicente, J. (2014). Lock-in or lock-out? How structural properties of knowledge networks affect regional resilience. *Journal of Economic Geography, 14*(1), 199–219.

Cross, R., Grinfield, M., & Lamba, H. (2009). Hysteresis and economics. *Control Systems IEEE, 29*(1), 30–43.

Davidson, D. J. (2010). The applicability of the concept of resilience to social systems: Some sources of optimism and nagging doubts. *Society and Natural Resources, 23*(12), 1135–1149.

Dawley, S. (2014). Creating new paths? Offshore wind, policy activism, and peripheral region development. *Economic Geography, 90*(1), 91–112.

Dawley, S., Pike, A., & Tomaney, J. (2010). Towards the resilient region? *Local Economy, 25*(8), 650–667.

Feldman, M. P. (2007). Perspective on entrepreneurship and cluster formation: Biotechnology in the US Capitol region. In K. R. Polenske (Ed.), *The economic geography of innovation* (pp. 241–260). Cambridge University Press.

Folke, C. (2006). Resilience: The emergence of a perspective for social-ecological systems analyses. *Global Environmental Change, 16*(3), 253–267.

Foster, K. A. (2007). A case study approach to understanding regional resilience, Working Paper, Institute Urban and Regional Development, UC Berkeley.

Frenken, K., Van Oort, F., & Verburg, T. (2007). Related variety, unrelated variety and regional economic growth. *Regional Studies, 41*(5), 685–697.

Giannakis, E., & Bruggeman, A. (2017). Determinants of regional resilience to economic crisis: A European perspective. *European Planning Studies, 25*(8), 1394–1415.

Gong, H., & Hassink, R. (2017). Regional resilience: The critique revisited. In N. Williams & T. Vorley (Eds.), *Creating resilient economies: Entrepreneurship* (pp. 206–216). Edward Elgar.

Grabher, G. (1995). The disembedded regional economy: The transformation of East German industrial complexes into Western enclaves. In A. Amin & N. Thrift (Eds.), *Globalization* (pp. 177–195). Oxford University Press.

Grabher, G., & Stark, D. (1997). Organizing diversity: Evolutionary theory, network analysis and post-socialism. *Regional Studies, 31*(5), 533–544.

Grabher, G. (1993). The weakness of strong ties: the lock-in of regional development in the Ruhr area. In G. Grabher (Ed.), *The embedded firm: on the socioeconomics of industrial networks* (pp. pp. 255–277). Routledge.

Hassink, R. (2010a). Regional resilience: A promising concept to explain differences in regional economic adaptability? *Cambridge Journal of Regions, Economy and Society, 3*(1), 45–58.

Hassink, R. (2010b). Locked in decline? On the role of regional lock-ins in old industrial areas. In R. Boschma & R. Martin (Eds.), *The handbook of evolutionary economic geography* (pp. 450–468). Edward Elgar.

Hassink, R. (2017). Advancing the understanding of regional economic adaptability in a non-western context: An introduction to the special issue. *Growth and Change, 48*(2), 194–200.

Hill, E., Wial, H., & Wolman, H. (2008). Exploring Regional Economic Resilience, Working Paper, Institute Urban and Regional Development, UC Berkeley.

Hu, X. (2014). State-led path creation in China's rustbelt: The case of Fuxin. *Regional Studies, Regional Science, 1*(1), 294–300.

Hu, X. (2017). From coal-mining to coal-chemicals? Unpacking new path creation in an old industrial region of transitional China. *Growth and Change, 48*(2), 233–245.

Hu, X., & Hassink, R. (2017b). Place leadership with Chinese characteristics? A case study of the Zaozhuang coal-mining region in transition. *Regional Studies, 51*(2), 224–234.

Hu, X., & Hassink, R. (2017c). New perspectives on restructuring of old industrial areas in China: A critical review and research agenda. *Chinese Geographical Science, 27*(1), 110–122.

Hu, X., & Hassink, R. (2017a) Exploring adaptation and adaptability in uneven economic resilience: a tale of two Chinese mining regions. *Cambridge Journal of Regions, Economy and Society* (forthcoming).

Ibert, O., & Schmidt, S. (2014). Once you are in you might need to get out: Adaptation and adaptability in volatile labor markets—the case of musical actors. *Social Sciences, 3*(1), 1–23.

Isaksen, A. (2015). Industrial development in thin regions: Trapped in path extension? *Journal of Economic Geography, 15*(3), 585–600.

Isaksen, A., & Trippl, M. (2016). Exogenously led and policy-supported new path development in peripheral regions: Analytical and synthetic routes. *Economic Geography*. https://doi.org/10.1080/00130095.2016.1154443

Kirsh, D. (1996). Adapting the environment instead of oneself. *Adaptive Behavior, 4*(4), 415–452.

MacKinnon, D. (2011). Beyond strategic coupling: Reassessing the firm-region nexus in global production networks. *Journal of Economic Geography, 12*(1), 227–245.

MacKinnon, D., Cumbers, A., Pike, A., Birch, K., & McMaster, R. (2009). Evolution in economic geography: Institutions, political economy, and adaptation. *Economic Geography, 85*(2), 129–150.

MacKinnon, D., & Derickson, K. D. (2013). From resilience to resourcefulness: A critique of resilience policy and activism. *Progress in Human Geography, 37*(2), 253–270.

Markusen, A. (1999). Fuzzy concepts, scanty evidence, policy distance: The case for rigour and policy relevance in critical regional studies. *Regional Studies, 33*(1), 869–884.

Martin, R. (2012). Regional economic resilience, hysteresis and recessionary shocks. *Journal of Economic Geography, 12*(1), 1–32.

Martin, R., & Sunley, P. (2006). Path dependence and regional economic evolution. *Journal of Economic Geography, 6*(4), 395–437.

Martin, R., & Sunley, P. (2015). On the notion of regional economic resilience: Conceptualization and explanation. *Journal of Economic Geography, 15*(1), 1–42.

Martin, R., Sunley, P., Gardiner, B., & Tyler, P. (2016). How regions react to recessions: Resilience and the role of economic structure. *Regional Studies, 50*(4), 561–585.

Martin, R., & Sunley, P. (2017). Competitiveness and regional economic resilience. In R. Huggins & P. Thompson (Eds.), *Handbook of regions and competitiveness: contemporary theories and perspectives on economic development* (pp. 287–307). Edward Elgar.

Neffke, F., Henning, M., & Boschma, R. (2011). How do regions diversify over time? Industry relatedness and the development of new growth paths in regions. *Economic Geography, 87*(3), 237–265.

Nyström, K. (2017). Regional resilience to displacements. *Regional Studies*. https://doi.org/10.1080/00343404.2016.1262944

Oliva, S., & Lazzeretti, L. (2017). Adaptation, adaptability and resilience: The recovery of Kobe after the Great Hanshin Earthquake of 1995. *European Planning Studies, 25*(1), 67–87.

Peck, J., & Theodore, N. (Eds.). (2015). *Fast policy: Experimental statecraft at the thresholds of neoliberalism*. University of Minnesota Press.

Pendall, R., Foster, K. A., & Cowell, M. (2010). Resilience and regions: Building understanding of the metaphor. *Cambridge Journal of Regions, Economy and Society, 3*(1), 71–84.

Pike, A., Dawley, S., & Tomaney, J. (2010). Resilience, adaptation and adaptability. *Cambridge Journal of Regions, Economy and Society, 3*(1), 59–70.

Robinson, G. M., & Carson, D. A. (2015). Resilient communities: Transitions, pathways and resourcefulness. *The Geographical Journal, 182*(2), 1049–1061.

Sayer, A. (1991). Behind the locality debate: Deconstructing geography's dualisms. *Environment and Planning A, 23*(2), 283–308.

Simmie, J., & Martin, R. (2010). The economic resilience of regions: Towards an evolutionary approach. *Cambridge Journal of Regions, Economy and Society, 3*(1), 27–43.

Smit, B., Burton, I., Klein, R., & Wandel, J. (2000). An anatomy of adaptation to climate change and variability. *Climate Change, 45*(1), 223–251.

Tomlinson, P. R., & Branston, J. R. (2014). Turning the tide: Prospects for an industrial renaissance in the North Staffordshire ceramics industrial district. *Cambridge Journal of Regions, Economy and Society, 7*(3), 487–507.

Chapter 3
The Economic Resilience of Old Industrial Cities in China with Different Types of Crises: A Call for Multiple Perspectives

3.1 Introduction: Why Chinese OICs and Their Resilience Matter?

In the Chinese context, the theme on economic resilience of old industrial cities (OICs) is not a new topic. Notably since 2003, this topic has been receiving much attention both from Chinese economic geographers, policy-makers and the media. This research theme, indeed, is largely stimulated by the national policy of 'Revitalize Northeast China' in 2003 and, more recently, by the top-down financial aid program of 'Transform Resources-exhausted Cities' since 2008. The two policy programmes are aimed to help OICs to overcome economic stagnation and decline under the external (also internal) contextual changes, such as marketization, globalization and resource exhaustion. to be more specific, although most studies are policy-oriented rather than process-focused, China's OICs have been generally recognized as typical 'problem regions' which ought to be at the core of China's regional development research (Lu & Fan, 2010). On the one hand, many OICs still suffer from long-term sluggish growth or even stagnation due to ill-balanced economic and social structures inherited from the planned economy. On the other hand, these regional economies, characterized by a high degree of industrial monopolization and specificity, are vulnerable to all sorts of shocks in a transitional context, particularly when it comes to the changes of national strategies and institutions. Recently, the slowdown of China's economy and the increasing emphasis on the quality rather than quantity of economic development have put OICs in a dilemma: how to shift from the existing resource-based, highly contaminated and state-dominated regional economies to more market-oriented, green and sustainable ones, but in the meanwhile without being harmful to local economic growth and social harmony. That is to

(This chapter is adapted from and revised based on: Hu, X., & Hassink, R. (2017). New perspectives on the restructuring of old industrial areas in China: A critical review and research agenda. Chinese Geographical Science, 27(2): 110–122.)

X. Hu, *Uneven Economic Resilience of Old Industrial Cities in China*,
Urban Sustainability, https://doi.org/10.1007/978-981-99-9279-9_3

say, the top-down policy pressure, combined with very specific and grounded assets, agencies and crises OICs have, can lead to potentially divergent processes, mechanisms and abilities of regional economic resilience. Moreover, compared to other types of regions in China, OICs with their industry over-specialization and locked-in structures are more exposed to external shocks and more economically vulnerable in crises. In a theoretical sense, doing in-depth research on economic resilience of OICs can contribute to the understanding of varieties of regional development 'models' in China (Hu, 2017; Wang et al., 2014; Zhang & Peck, 2014).

Although this topic is relevant in both policy and academic circles, existing research on it is weak. A main weakness is the neglect of using modern, sound and refreshing concepts to systematically analyze and explain the evolutionary nature of OICs and, thus it fails to theorize Chinese practices and contribute to the international literature. This pitfall, indeed, is ascribed mainly to the Chinese distinct 'practice-based' characteristic of economic geographic research. That is, as some leading Chinese geographers critically reflected (Liu & Lu, 2002; Liu et al., 2011), the primary target and driving force of the discipline development are to fulfill the national policy demand. As a result, many issues with regard to the restructuring of China's OICs remain poorly understood. For instance, little research has been devoted to unveil how OICs evolve and adapt through time in an on-going market reforming and uncertain yet still globalizing context. Who plays what role in affecting the processes and for what purpose? And why are some OICs more economically resilient while others are not? These questions, in our view, are crucial to understand the uneven development and resilience of OICs in different but grounded contexts of shocks/crises, and need to be equipped by sound concepts and perspectives in a more holistic way. As suggested by Liu (2009), it is badly needed for geographers to (critically) adopt mainstream theories to rethink highly contextualized Chinese practices, as a useful channel to launch an intelligent 'bridge' with mainstream economic geography.

Globally, there has been a notable paradigm shift within the realm of economic geography, transitioning from conventional neoclassical viewpoints to a more heterodox framework rooted in institutional, relational, and evolutionary perspectives (Hassink et al., 2014). This shift is marked by two distinct yet interconnected trajectories. On one hand, a range of evolutionary principles such as path dependence, lock-in, path creation/development, and concepts like (un)related variety, along with innovative metaphors such as resilience, have been employed to dissect the resurgence and decline of Older Industrial Cities (OICs) across Europe and North America (Fornahl et al., 2012; Grabher, 1993; Hassink, 2010; Simmie & Martin, 2010). On the other hand, alternative perspectives provide supplementary explanations that complement these Evolutionary Economic Geography (EEG) approaches.

In this vein, it has been argued that the role of the state, particularly in terms of governance and policies intertwined within a dynamic and multi-tiered political economy, plays a pivotal role in shaping the industrial landscape of Older Industrial Cities (OICs) (Dawley, 2014; Hudson, 2005; Hu, 2014). Others contend that incorporating the agency of individuals into analyses can enhance the comprehension of micro-level dynamics and how they collectively contribute to adaptation and adaptability (Bailey et al., 2010; MacNeil & Steiner, 2010). Collectively, these concepts have demonstrated substantial explanatory capacity in deciphering the intricate evolutionary trajectory of OICs.

However, it's noteworthy that few, if any, of the abovementioned concepts have been explicitly employed in the context of research concerning Chinese OICs, signifying a critical research gap that warrants attention. As such, this chapter embarks on the exploration of potential avenues for regional economic resilience research on China's OICs. The central focus lies in identifying international concepts applicable to comprehending the evolution and spatial dynamics of OICs within China. Drawing upon a thorough and critical review of literature in both English and Chinese, this chapter proposes a series of perspectives and reframes an EEG and REG combined conceptual framework of regional economic resilience . I also argue that the economic resilience of Chinese OICs and its uneven nature over space and time is not only shaped by conventional economic crises, but also affected by the likely long-term aftershocks of the COVID-19 pandemic.

3.2 Old Industrial Cities in China: History and New Developmental Trends

China's OICs include natural resource-based (e.g., mining, forestry, petroleum) and heavy industrial (steel, iron, machinery and shipbuilding, etc.) territorial complexes, as well as regions specialized in light industries (e.g., textile), which were developed under the 1949–1976 centralist regime and gradually became economically stagnated in the course of market reforms. Differed from many Western OICs whose fortunes are closely connected to market competition and technological dynamics, China's OICs have been evolving in a distinctive manner. Their evolutionary trajectories seem to be more influenced by national leadership and policy dynamics (Ding & Sun, 2006; Lin, 2002; Wei, 2000). The development history of OICs in China can be divided into three major phases given the shifts of national government leadership, namely growth in post-1949 centralization era, stagnation in post-reform era (1979–2002), and restructuring in the 2003–2013 period (Table 3.1).

(1) *Emergence and growth under centralization era (1949–1976)*

Since 1949, to consolidate the central leadership and the new government after the civil war, a considerable number of military-related industries were initiated in the coastal region under a defense-centered ideology. Meanwhile, the pro-Soviet national-state also prioritized to industrialize the northern border regions, which were endowed with plentiful mineral resources. The industrialization of Northeast China, for example, was primarily empowered by the First and Second Five-year Plan (第一和第二个五年计划) (1953–1962), in which China as a new member of

Table 3.1 National changing policies and their impact on OICs in China

Phase	Leadership era and its OIC related policies	Major impact on OICs
Expansion and growth	Mao's centralization era (1949–1976): • The 1st and 2nd Five-Year Plan • The Great Leap Forward • The Third Front Construction	Northeast became China's most industrialized region • A variety of military-specialized regions and natural resource-based cities and districts were extensively established in inland China
Stagnation and decline	Deng's neoliberalism and Zhu's radical SOEs reform era (1979–2002): Rural privatization • Costal centered policy • Radical Reform on SOEs • Go West	• The rapid growth of coastal regions at the cost of the socio-economic decline of OICs • The dramatic decrease of SOEs in market share. SOE-dominated OICs suffered from economic stagnation and social unrest • Western OICs started to receive central investments
Revitalization and revival	The Hu-Wen administration (2003–2013): • Revitalizing the Northeast • The Rise of Central China • Transforming Resource-exhausted Cities	• Northeast China underwent a fast-growing under national policy-led inputs • OICs in central China benefited from supports for industrial upgrading • A total of 69 resource-based cities received top-down financial aids to foster regional industrial transformation

Source Author's compilation based on official policy documents

the communist alliance received numerous free investment projects (e.g., heavy, power and manufacturing) from the former Soviet Union (Zhang, 2008). Likewise, OICs in Southwest China can be traced back to the national radical defensive policy triggered by the Sino-Soviet split, namely the Third Front Construction (三线建设) (1964–mid 1970s). In order to preserve military strengths and and heavy manufacturing capacities for a potential war with the former Soviet Union, the central government undertook a large-scalar industrial transplantation, in which almost all military-related sectors, together with numerous workers in the coastal belt were mandatorily relocated to the western hinterland (Naughton, 1988). Consequently, this movement led to a large spatial expansion of heavy and mining industries, which outlined the general profile of China's OICs (Fig. 3.1).

(B) *Stagnation and decline in reform era (1979–2002)*

Since 1979, however, the formerly prosperous areas began to experience a painful stagnation. This are explained by three inter-linked reasons: First, the SOE-based regional economies, in which planned institutions, corporate bureaucracies and entrenched territorial input–output production systems were well preserved,

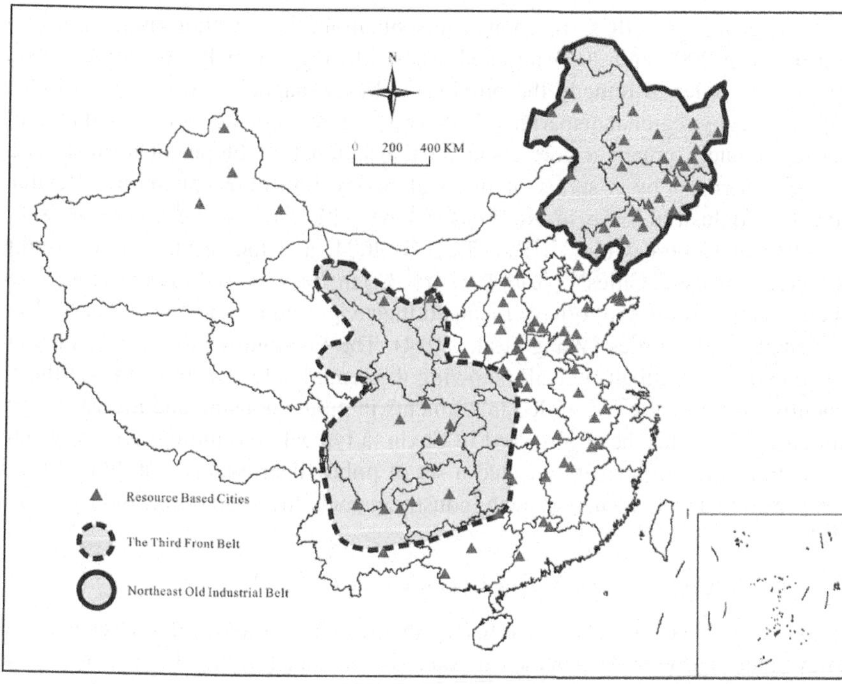

Fig. 3.1 Geographical Distribution of China's old industrial areas. *Source* authors' design

presented many difficulties and even resistance to market reforms. Secondly, the marketization reform facilitating private and collective economies squeezed the market share of SOEs and thus negatively impacted on OICs. Besides, the 'Coastal Area Preferential' (沿海地区优先) neoliberal policy broke the post-1949 egalitarianism for regional development, leading to a spatially unbalanced development (Dunford & Li, 2010). This 'benefiting some places but punishing others' policy, can be used to explain why industrial complexes in Northeast and Southwest China lost dynamics and mutated into less-favored OICs (Wang et al., 2014). Thirdly, in 1998, the government tightened up the policies on SOE privatization by introducing so-called 'modern enterprise institutions' (现代公司制度). In a short run, firm ownership, institutions and size were variegated, redesigned and diminished, respectively (Garnaut et al., 2006), which directly caused massive factory closures and economic downturn in OICs. Although the central state became aware of the increasing development gap between the coast belt and inland OICs and tried to bridge it by launching the 'Western Development' (西部大开发) strategy in 2000, OICs still constrained to develop new industries.

(C) *Revitalization since restructuring period (2003–2013)*

It is fair to say that the top-down neoliberalism and post-1998 radical SOE reforms relaxed the development of capitalism in coastal China. However, it is at the expense of the stagnation of OICs. To combat this situation, the fourth-generation leadership between 2003 and 2013 initiated a new ideology, namely 'Scientific Development', in order to remedy the previous policy's inappropriateness and to alleviate increasing regional disparities. In this respect, revitalizing less-favored regions was legitimately put as a core issue on the national development agenda. This can be evidenced by a series of national policy portfolios, including 'Revitalizing Old Industrial Base in Northeast China' (振兴东北老工业基地) in 2003, the 'Rise of Central China' (中部崛起) in 2004, and the recent 'Transforming Resource-exhausted Cities'[1] (资源型城市转型) in 2008. With these strong national inputs, many OICs seem to have regained their vigor and some have embarked on rejuvenation trajectories (Wang et al., 2014). The Economist (2012) published an article entitled 'Rustbelt revival', showing the remarkable growth of the northeast benefiting from massive state-led investments in public housing and infrastructure. Similar evidence has been observed in Fuxin, a typical coal-mining region, where the central state purposively enacted a set of policy instruments that ignited local enthusiasm to develop a new growth industry, namely the wind power industry (Hu, 2014).

(D) *New development trends in new contexts (2013–now)*

Although GDP in China's OICs has been growing under the above-described national policy inputs, it does not mean they are successfully transforming. In fact, the policy-driven economic growth model generates an illusion that covers up the weak endogenous capacities in OICs. This can be exemplified by the recent figures on GDP growth in the first half of 2014 (NBSC, 2014), showing that many mature industry-based provinces, such as Shanxi, Hebei, Heilongjiang, Jilin, ranked in the bottom of China's 31 provinces. The economic situation is particularly bad in the northeast, where old command-economy institutions dominate, the central-planning mentality of people persists, and young people are more likely to emigrate elsewhere for better jobs (Economist, 2015a). The recent downturn of GDP growth in OICs is partly caused by that the so-called 'New Normal' national strategy (OECD, 2015). In this framework, new policies and regulations in pursuit of 'structural change' and 'innovation driven development model' are being enforced. In this sense, OICs are in a tough position for two reasons. First, it is difficult to change the dominant, geographically and historically well conditioned 'resource and investment-led development model' in OICs. Secondly, when it comes to reforms, local actors are exercised about the potential economic loss and social insecurity that highly matter their political future.

Of course, the new normal state context can also provide opportunities such as new incentives, institutions and preferential policies for OICs. A set of top-down industrial policies shortly called 'amalgamation and recombination' (兼并重组), for example, was released since 2008. It aims at maintaining the competitiveness of OICs in the post-crisis context through fostering the spatial restructuring by cross-regional/hierarchical mergers, agglomerations and acquisitions of SOEs. Additionally, large SOEs are incentivized to implement the so-called 'going-out' strategy. By doing so, domestic old industries can be transplanted to other emerging economies and/or absorbing foreign technologies and natural resources—as a way to revitalize SOEs

(Moody, 2013). In particular, an accelerated internationalization of Chinese SOEs can be expected under the Belt and Road Initiative, which will be further integrated local economies of OICs into global production networks and international markets, and thus be shaped in the future. The other way around, the global crisis has also accelerated a new round of global shift of traditional sectors from industrialized economies to China, which to some extent facilitate the integration of China's OICs into global production networks that may lead to regional renewal (Hudson & Swandon, 2012).

Recently, the "Report on the 20th National Congress of the Communist Party of China" (二十大) underscores the imperative to foster regional harmonization and provide targeted support for the advancement of "Special Category Regions" (特殊类型地区, SCRs) as pivotal strategies for addressing the prevailing issues of imbalanced and insufficient development in our country. SCRs encompass underdeveloped areas, historically significant regions, border zones, ethnically diverse territories, ecologically fragile zones, resource-based cities, and OICs. Particularly, negatively affected by the COVID-19 crisis and the overall turn-down trend of China's marcoecomic development. These regions, indeed, have confronted distinct challenges, yet, under China's Regional Balanced Development Strategy (区域协调发展战略), concurrently constitute "crucial focal points for propelling high-quality development and undertaking distinctive roles. Compared to China's coastal, metropolitan core regions, whose economies are characterized by diversified industries, prolific knowledge establishments, and permissive institutional frameworks, SCRs often bear the negative historical burden of socio-economic development, manifest signs of structural lock-ins, and confront many institutional and social hurdles in nurturing new industries and technologies. They epitomize quintessential problematic and marginalized territories, necessitating commensurate dispensation of "region-specific solicitude" (Li, 2009, 2022; Zhang et al., 2016). In general, the majority of special-category regions continue to lag behind within the national economic spatial framework, confronting longstanding practical challenges of "difficulty in transitioning from old industries, establishing new industries, and embedding these new industries in the local." (旧产业难转、新产业难立、立了难扎根). In this sense, the exploration of mechanisms for shaping novel industrial trajectories in China's SCRs (including OICs) has emerged as a central proposition in the ongoing research theme in China's economic geography (Hu et al., 2020).

To summarize, China's OICs have four key characteristics: (1) geographically, they are mainly located in Northeast and Southwest China, and also consist of a considerable number of resource-based cities in Central China. (2) These regions are characterized by a multitude of distinct types, and each might possess unique attributes and developmental logics. Significantly divergent characteristics exist not only among different categories but even within the confines of the same category. (3) The current mature industrial space is a result of a combination of regional endowments, historical events and policy dynamics. (4) Recently, OICs seem to be increasingly affected by national state-led rescaling and economic globalization, although local actors play a role.

3.3 OIC Research with Chinese Characteristics? A Critical Review

In our quest to elaborate on the existing understanding of the restructuring of OICs in China, we think Chinese-speaking journals could provide important input. We therefore selected six major Chinese-speaking geography journals, namely 'Acta Geographia Sinica', 'Scientia Geographica Sinica', 'Progress in Geography', 'Economic Geography', 'Areal Research and Development' and 'Human Geography'. We found and critically reviewed forty-three articles in these journals which referred to 'OICs' (such as 'old industrial bases' and 'resources-exhausted cities') and 'transformation or restructuring' in title or keywords.

One research strand is oriented to explaining why OICs in China stagnated (Han et al., 2005; Li & Li, 1996; Sun & Ding, 2005). These studies mainly comprise of analyses of the constraining role of industrial-sectorial impact factors such as the power incoherence between centrally controlled SOEs and self-contented local states, the lack of regional integration, the specialization 'trap', etc. Other contributions have highlighted the salient role of historically inherited institutions in explaining the decline of OICs. According to Ding and Sun (2006) the real problem of OICs is "not economic, but an institutional problem" (p. 1). They argued that the strong resistance to change in OICs was fundamentally ascribed to inappropriate national institutional arrangements, such as top-down controls and unfair tax policies. Xiao and Mao (2008), for instance, stressed that the stagnation of the northeast stems from the duplicate of the Soviet-style model of 'Territorial Industrial Complex'. Despite some advantages at the beginning stage of internationalization, this model, for a long term, is not favorable and competitive in a market economy. Because the model's strong focus on the encompassing effects of territorially bounded and self-sufficient production systems has led to a prevailing social-industrial system in China. This system resulted in many rigid and harmful institutions, such as the idea of '*whatever size but being complete*'[2](不求大小, 但求完整), and the '*danwei*'(单位)[3] institution. Some researchers view these institutions as a '*culture hindrance*'(文化障碍) that can have a long-term negative impact on OICs (Dong & Zhang, 2008; Wang & Xiao, 2004). In their studies, the danwei institution, for example, was defined as a cognitive barrier that avoid local actors from having new career visions, as working in a danwei stands for a symbol of higher social status and well-being. Zhang (2006) assumed that even if this institution was removed, the feelings of the '*iron rice bowl*'(铁饭碗)[4] can be passed on from older to younger generations, which consistently play a role in undermining entrepreneurship and creativity in OICs.

In summary, while these contributions have addressed a variety of constraining impact factors, particularly the planned institutional 'inertias' that led to the stagnation of OICs, they are too general and lack thorough empirical evidence. Although history seems to have played an important role, it is still unclear how institutions have been embedded into place distinctive contexts, and why some regions have tended to stick stronger to old routines, networks and structures while others do less. Moreover, the studies strongly rely on a top-down, downward line of arguments, by emphasizing that it is top institutional designs that made OICs stagnant, but little attention has been paid to upward interactive processes of certain regions with old institutions. The people in metropolitan OICs, for instance, tend to find new jobs not limited to SOEs, as large cities can offer more diversified job opportunities. The evidence in Shenyang has shown that inner-city labor mobility between SOEs and non-SOEs can reshape people's mindset on employment and thus affect the evolution of OICs (Dong & Zhang, 2008; Hu et al., 2020).

Another research line deals with policy advices for directing the near-future restructuring of OICs in China (Chen et al., 2006; Wang et al., 2011; Zhang, 2004). Although most advises are derived from individual cases, they have pretty much in common. Most scholars claim that the restructuring of OICs is about either the extension of existing industrial chains or the recombination of local existing assets for new options. It has been argued, for example, that single-industry specialized regions, such as the forestry city Yichun in Heilongjiang Province and the petroleum city Kelamayi in Xinjiang Province, have little room to create entirely new industries, but can facilitate organizational and technological neighboring sectors related to the dominant industry (Wang et al., 2011; Zang et al., 2006). Similarly, based on a case study of Huangshi's tourism industry, Zhu et al. (2002) state that new growth paths in OICs can be fostered by exploiting and commercializing place-specific uncommitted assets. Zhang et al. (2011)'s insightful contribution, however, suggests that developing new industries or extending existing ones is contingent on localized conditions and different types of OICs. Another common focus is given to the role of the state. Given the economic devolution to the local state, much literature regards the local state as the 'initiator' and 'driving force' for industrial change, suggesting that the local state can take effective actions to make OICs dynamic, such as through industrial relocation, tax reduction and land commercialization, etc. (Chen et al., 2006; Mao et al., 2008). In contrast to most policy-oriented studies, one process-focused contribution illustrated how local state entrepreneurism in the Ma'anshan steel city of Anhui Province facilitated local experimentations that significantly restructured the institutional base, which led to industrial diversification (Yin et al., 2010). Others, however, argued that the central state, rather than the local state should bear the brunt to revitalize OICs, as its past policy failures still negatively affect today's OICs (Ding & Sun, 2006; Xiao and Mao, 2008).

The latter research stream explicitly resonates with Liu and Lu's (2002) argument that "...Chinese economic geographers have been involved in many consulting jobs for the Chinese government" (p. 2123). In many studies, too much attention has been paid to analyzing general problems and drawing quick-fix policy recommendations. Despite much empirical research, the analyses mostly draw on secondary data rather than in-depth fieldwork-based primary data, which we consider as problematic particularly concerning the effects on the quality of arguments. Moreover, they lack a sound conceptualization, which means drawing empirical evidence for theorization is scant, and thus hardly yield general lessons. Some scholars highlight the salient role of the state in the restructuring of OICs, implicitly suggesting a political economy approach to rethink the evolution of OICs in China (see Ding & Sun, 2008). However, by far, these studies lack a dynamic, relational and multi-scalar account, both empirically and conceptually, and pay little attention to the restructuring processes and mechanisms of OICs. In reality, no region evolves in a purely linear 'top-down' or 'bottom-up' way, there are always multi-agencies and multiple spatial levels involved.

To sum up, it is concluded that despite several useful contributions the existing accounts have shown some clear deficits. The existing literature well echoes Liu's (2009) critique on the nature of economic geography in China being 'policy-oriented', 'pragmatic' and 'lacking theorizing', or as Coe et al. (2013, p. 512) have stated: "...Chinese economic geographers have tended to play the role of planners and consultants to state authorities rather than critical social scientists". Although a variety of building blocks, such as historical institutions, SOEs and state agency, are all highlighted; it is fair to say that nearly none of them are linked to corresponding international modern theories. The major weaknesses revolve around the limited concern for interactions between firms, institutions, policies and places, as well as for the underlying processes of key multi-scalar actors (e.g., individuals and agencies). Another explicit deficit can be seen in the fact that too much work is based on single-case studies, with little concern for the spatial differences in the restructuring of OICs and the underlying causes and outcomes.

Furthermore, recent reflections from several key Chinese economic geographers on the topic have shown that mainstream theories have been constructed based on the advantages of Western regions, often encountering explanatory bottlenecks when applied to special-category regions (He & Zhu, 2020). However, a considerable body of evidence in reality demonstrates that special-category regions not only give rise to new industries but also exhibit heterogeneity in the mechanisms and effects of such shaping, leading to instances of '(Western) theory lagging behind (Chinese) reality' (Qiu et al., 2018; Zhang & Yu, 2023; Zhou & He, 2019). Evidently, for special-category regions, regional structural endowments do not represent deterministic factors nor are they purely restrictive elements; rather, agency plays a significant role. Consequently, new research questions have emerged, such as, "through what interactive mechanisms do structure and agency shape regional trajectories of new industries?"; "Do differences exist in terms of characteristics, interaction patterns, and outcomes between the two in different regions?" (Grillitsch & Sotarauta, 2020). These questions challenge existing theoretical frameworks and necessitate a more dialectical approach to the interaction between structure and agency. Furthermore, there is an urgent need to transcend research biases favoring Western core regions.

3.4 Modern Concepts and Frameworks on the Economic Resilience of China's OICs

Based on the current state of OICs restructuring-related international literature, we identify five key yet not mutually exclusive, themes with a particular focus on the concepts used in dealing with the dynamics of OICs in Europe and North America.

The first research theme is centered on why OICs lose dynamics and fail to grow further. Initially, some theories from economics are applied to examine the mechanism of economic decline in OICs, such as cumulative causation, long waves, over-specialization, and the product life cycle (Hassink & Shin, 2005). Many of these theories were developed out of criticism towards the neoclassical theory of regional development, which emphasize the equilibrium function of capitalism. Many of the problems in the OICs seemed to be structural and were hence not solved by the invisible hand of the market economy, as has been suggested by neoclassical economists.

Secondly, later on studies started to concentrate more on the dynamics and mechanisms of the restructuring of OICs by using the evolutionary concepts of path dependence and lock-ins. "A path-dependent process or system is one whose outcome evolves as a consequence of the process' or system's own history" (Martin & Sunley, 2006, p. 399), whereas Martin (2010, p. 3) defines lock-ins as the notion that "…most fully captures the idea that the combination of historical contingency and the emergence of self-reinforcing effects, steers a technology, industry or regional economy along one 'path' rather than another". Closely related to the discussion around path dependence and regional evolution is the issue of lock-ins hindering necessary restructuring processes in regional economies (Hassink, 2010). Lock-in situations appear because specialized industrial regions endowed with particular resources, competences and institutional structures are unable to match changing market requirements. Grabher's study (Grabher, 1993) on the Ruhr area, Germany, uncovered that over-specialization is highly embedded in a place-specific environment, in which local inter-firm nexus, political frameworks, and key actors' cognitive mindsets are well developed and reinforced over time that locked in a hard-to-change situation. Building upon earlier analysis on (negative) path dependence from evolutionary economics (Arthur, 1994; David, 1985), his work has inspired many researchers to take history seriously in explaining the long-run stagnation of OICs (Boschma & Lambooy, 1999; Hassink, 2010; Hassink & Shin, 2005). Political lock-ins, in particular, as a constraining impact factor that not only hinders novelties entering into old industries but also prevents local agents to interactively learn from elsewhere (Cho & Hassink, 2009; Hassink, 2010; Hassink & Shin, 2005; Hudson, 2005). Hassink (2007), in contrast, illustrated how the textile region of Westmünsterland in Germany took advantage of weak institutional lock-ins to foster industrial renewal.

Thirdly, the concepts of path creation and industrial relatedness are used to demonstrate how (pre) existing technologies, skills, networks and institutions are transferred to contribute to the emergence of new industries in a region. Path creation gives a historical view of how path contingency and place dependence co-evolve through time to generate 'creative destruction' that brings new developmental impetus. Evidence related to path creation can be found in many OICs across the world: the new compound materials sector emerged out of the metal industry in Styria, Austria (Tödtling & Trippl, 2004), the microelectronics industry was built on the watch manufacturing in Jura of Switzerland (Maillat et al., 1997), and the formation of low-carbon technologies benefited from the pollution reduction activities in the steel and iron industry in China (Cooke, 2011). More recently, newly developed concepts in the evolutionary economic geography (EEG) literature, such as related variety (Frenken et al., 2007) and regional branching (Boschma & Frenken, 2011), emphasize the role of 'technological relatedness' in facilitating inter-firm/industry learning, technological spillovers and knowledge combinations that lead to innovation, job creation and diversification. Increasing evidence has shown that new industries are tended to branch from existing technologically proximate sectors (Boschma et al., 2013; Neffke et al., 2011).

Fourthly, EEG has recently been criticized for its narrow firm-based perspective and methodological reductionism (Martin, 2010, 2012). In a special issue in *Economic Geography*, it is suggested to engage evolutionary thoughts into broader institutional contexts, in general, and to integrate a geographical political economy approach into EEG, in particular (MacKinnon et al., 2009). Some evidence has shown that path de-locking or creation in OICs are not just about firm-based market-driven processes, but are rather influenced by extra-local, higher-level systems and actors, such as the state, politics and policies in multi-layered and changing socio-institutional environment (Dawley, 2014; Hudson, 2005; Hu, 2015). Another complementary perspective to EEG approach is an agency view that rejects seeing firms as a whole behavioral agent but instead pays more attention to the role of human individuals or groups of individuals in the change of OICs. By adopting the notion of place leadership, for example, MacNeill and Steiner (201) emphasized the major influence of the Styrian Minister Herbert Paierl on the transformation of Styria in Austria. Similar evidence can be also found in Bailey et al. (2010)'s study, in which the original sources for industrial renewal of the West Midlands, UK and Tuscany, Italy were derived from key political leaders.

Fifthly and strongly related to some above-mentioned concepts, recent research focuses on explaining why some OICs are able to adapt more positively and move onto new paths whereas others remain locked in decline/slow growth over time. In contrast to other research themes focusing on single cases by single approaches, this theme aims at seeking an open-ended, multi-disciplinary and pluralist approach to explore the uneven restructuring of OICs. By introducing a comparative framework, several impact factors affecting the restructuring of OICs are identified, such as varieties of capitalism among nations (Birch et al., 2010), distinct degrees/types of political-institutional lock-ins which can lead to different regional adaptability (Hassink, 2010), and different configurations of place leadership and social capital which enable the (re)framing of (new) organizations and institutions in some OICs but not in others (Bailey et al., 2010; Safford, 2009).

Based on the critical review in Sect. 3.3 and the overview of internationally recognized concepts in this section, Table 3.2 outlines which modern concepts might be used to overcome the three identified research deficits. The next section will explain in more detail how exactly the modern concepts can overcome existing research deficits and what kind of research questions can be derived from linking the deficits with the modern concepts.

Table 3.2 Modern concepts for overcoming main research deficits

Key questions	Existing deficits	Modern concepts (perspective)
• What impact factors cause OIC to stagnate or make OICs dynamic to change?	• Focus on the constraining role of history but without in-depth accounts • Lacking concern for place distinctive history and its causal mechanisms	• An evolutionary perspective • An institutional perspective • Path dependence and lock-in • Relatedness and path creation
• Who plays what kind of role in restructuring OICs in terms of new industry formation?	• Over-emphasis on the role of the state • Lacking multi-scalar & -agent views • Ignoring the role of individuals such as officials, entrepreneurs, stakeholders	• A human-agency perspective • A relational and multi-scalar view • Place leadership
• Why and how do OICs adapt differently leading to regionally uneven adaptability?	• Too many single case studies • Lacking knowledge of geographies of adaptation and adaptability	• A comparative perspective • A geographically uneven thinking

3.5 Engaging Regional Resilience Research in China's OICs: A Call for Multiple Perspectives and Agendas

(1) *EEG concepts and approaches*

Since Chinese economic geographers have recognized the salient role of history, particularly concerning the institutions derived from the previous regimes that still affect many OICs in China, the concepts of path dependence and lock-ins might be relevant, particularly when it comes to interpreting the difficulties for industrial change and economic resilience in an OIC (Table 3.1). For example, path dependence could be used to inform how previous *danwei* institutions and related organizational tissues are inherited and reproduced locally, and in what conditions old routines and habits are maintained that undermine the processes and abilities of regional resilience. A common policy recommendation for some OICs to survive is to relocate existing mature sectors to elsewhere, but this can be read as an action of path extension which, from a long-term perspective, might be subject to a regional lock-in, leading to a weak regional resilience. In this sense, Grabher's (1993) lock-in concept could be useful to help to identify the intrinsic sources and types of lock-ins in a certain OIC, and to examine how local and extra-local conditions play a role in the process of regional economic resilience.

The concept of path dependence and its related notion of path creation also provides a dialectical point of view to think to what extent place-specific existing assets can be recombined or transcended into a basis for new paths to emerge (Table 3.2). In fact, neither 'jiexu chanye' (continued industries, 接续产业) nor 'tidai chanye' (alternative industries, 替代产业) emerge from scratch; both of them draw more or less on regionally inherent assets, and are built on an existing institutional framework. A history-informed and place-based approach can allow policy-makers to see both local existing and latent assets as a source for regional change. It also helps identifying which industry has the potential to grow given the very specific context. The 'relatedness' conception, for example, is promising in soundly identifying whether it makes sense to foster existing or alternative industries, and then which particular sectors have most potential to become new growth paths, and then subsequently reshape the nature of economic resilience for a structural change (Hu & Hassink, 2017a). Adopting these concepts is thus essential for the formulation/implementation of locally embedded policies, which reject introducing policy panaceas merely based on some 'good practices' from elsewhere.

Based on the above discussions, some questions are proposed for future research: (1) how do place-based history and planned institutions generate (or not generate) path dependence mechanisms in OICs? (2) How do local actors read, interpret and deal with local existing industrial, institutional and network structures? (3) To what extent can the mechanism of new path creation in the Chinese context of OICs be linked to 'related variety' and 'regional branching'? And how to make sound, grounded policies for China's OICs by using EEG concepts?

(B) *IEG, the role of the state, and GPE*

Given the on-going impacts of changing institutional regimes and corresponding policy arrangements on OICs in China, and the recent criticism on the EEG framework for over-emphasizing endogenous dynamics for understanding resilience, we call a need to engage future research with a broader multi-scalar institutional perspective (Table 3.2). Two recent examples are presented as an illustration to support this need. First, Liu and Dunford (2010)'s study on the evolution of Chinese traditional medicine cluster in Tonghua moved beyond the narrow view of EEG, highlighting the key role China's macro institutional transitions played in enabling local corporate strategic behavior and cluster dynamics. It is suggested that industrial change of OICs in the Chinese context is more likely to be influenced by macro, systematic, and collective socio-institutional forces, despite local impact factors. Secondly, Hu (2015) explored the path creation mechanism of coal-chemicals in Zaozhuang, by which he revealed that the ranking system of SOEs in the region, which resulted in a fragmented and conflicted local political economy, prevented local coal firms starting business in the coal-chemical industry but only focusing on the old ways of doing things—mining. The new industry was largely initiated by external market-led entrepreneurial discovery process, rather than a result of local firm agency and local strategic desire from the state. The two examples suggest that the economic resilience of OICs in China tend to have been strongly affected by political-institutional impact factors and have shown a multi-scalar 'noisy' process. The role of both the local and national political economy, and the underlying power interplay among SOEs, private and public institutions from multiple scales should be placed at the core of research. This may be able to explain the contradiction with regard to the co-existence of multiple but incoherent institutions and paths, namely a region may present eagerness of maintaining old paths but simultaneously show interests in fostering alternative paths.

Therefore, the Chinese OIC context suggests taking the dynamic role of the state, policies and socio-political forces into account in addition to the firm-based perspective when studying economic resilience. In this respect, an encompassing approach that combines an institutional and multi-scalar perspective can be relevant to explain the complexity of evolution in OICs in China. There are three questions that need to be taken into analyses: (1) how do local, regional, national and global firms, stakeholders and authorities with a geographical political economy perspective change the local selection environment for new industry formation (or old industry exit) in Chinese OICs? (2) Related to the first question and, given the fact that unfinished decentralization (such as the *tiao-kuai* hierarchical administrative system and the ranking system of SOEs) is not uncommon in Chinese OICs, how does the institutional asymmetry and fragmentation affect local efficiency and quality for renewal? Precisely, how do local state actors interact with differentiated-ranking and multi-spatially embedded entities such as private firms, SOEs, public institutions and policy agencies, in combination, to influence regional economic resilience? (3) It is recommended to scrutinize the varieties of path-renewal/creation in terms of process, mechanism and outcome, through conducting cross-regional/national comparative research, so as to explore why some OICs more focus on developing continued industries while others are interested in initiating alternative ones.

(C) *Human agency and social perspectives*

The role of individuals is missing in the existing literature on the evolution of OICs in China. It is strongly suggested that future research needs to pay more attention to the role of human actors in general and local state officials in particular in the restructuring of OICs. As Ma (2002) has convincingly stated, "any conceptualization of local economic restructuring [in China] must humanize the state and take into consideration the role of the local leaders and their economic and political interests" (p. 1552). This is because in the Chinese context local government leaders are endowed with sufficient decision-making power to mobilize locally controlled natural and public resources for certain economic purposes. They are also politically encouraged to act as 'institutional entrepreneurs' seeking for regional economic growth, as long as the expected results do not come into conflict with national political and economic agendas/aims. This fuzzy boundary of states and enterprises has been apparent in the SOE-dominated OICs, where local officials may take formal or informal positions in SOEs. We thus plea for future research to examine the role of human-based state agency, such as local state leadership in shaping the evolution of OICs in China (Hu & Hassink, 2017b). Possible research questions can be (1) how and to what extent local government leaders and their leadership plays a role in facilitating regional resilience in term of ways in which actors and organizations adapt and take action for what/whom? (2) What is the causal relationship between local key elites, institutional change and the formation of new industries for what purpose? (3) How do state leaders in different OICs read the same shock, and how do they interpret and react to the changing contexts, and why they might act differently on regional economic resilience? We strongly assume that using a human agency perspective, especially the idea of place leadership can provide a better understanding of the motive and logic of micro-scalar institutionalization and bottom-up processes for enabling and forming regional economic resilience in China's OICs.

(D) *Contextual sensitivity and the nature of crisis*

Frist, and more importantly, research questions on regional economic resilience should be closely situated into the latest contexts in China. Particularly, in a post-crisis era Chinese OICs are more vulnerable and thus confront more difficulties than other types of regions in terms of economic recovery and redevelopment (Hu, 2012; Hu & Yang, 2019). We suggest that future research needs to think of the economic resilience of China's OICs. Attention should be paid not only on how OICs adaptively manage short-term recovery after somewhat one-off shocks, but also on how they act to cope with long-term adaptation for sustainable development, and why they adapt differently. One research avenue, for example, is to comparatively explore the adaptation and adaptability of different resource-exhausted cities in the face of a similar 'slow-burn' crisis (such as de-industrialization, resource depletion, populating ageing and shrinkage), and the determinants behind this process.

Second, regarding to the 'new normal state' (新常态) context, a recent article shows that despite the implementation of strict top-down environment measures,—through compelling localities to shut down heavily polluting factories—different regions may interpret these measures differently and implement them pragmatically according to their own interests (Economist, 2015b). This, indeed, mirrors a widespread bureaucratic phenomenon in China, namely '*shang you zhengce, xia you duice*' (上有政策，下有对策， namely, the top has measures while the bottom has countermeasures). Therefore, another research avenue might explore why some OICs take national policies seriously and put them into action while others not. We argue that it may be useful to apply a geographical political perspective, focusing not merely on localized conditions, but more on local-central relations and the broader changing institutional environment. Last but not least, putting China's OICs in transition into the changing global production networks, global value chains, and international labor mobility can also generate insightful findings.

Third, the study of regional economic resilience originated from the global financial crisis in 2008, giving rise to two mainstream perspectives: equilibrium thinking from economics and EEG theory. This reflects a shift from the notions of engineering resilience and ecological resilience to evolutionary resilience. In recent five years, within the macro backdrop of unprecedented changes in global political and economic relations, the rise of "globalization reverse" or "national protectionism", coupled with the impact of financial crises, natural and human-made disasters, and public health emergencies, pose severe challenges to the resilience of regional economies. The COVID-19 pandemic crisis since 2019, in particular, has tested the resilience of regional economies. The pandemic's impact on economic systems worldwide has been profound, challenging the ability of regions to withstand shocks, adapt, and recover. The disruptions caused by COVID-19 have highlighted both the strengths and vulnerabilities of regional economic resilience (Hu et al., 2022).

The COVID-19 crisis has prompted a reevaluation of what it means for a region to be resilient (Liu, 2020). It's not just about bouncing back from shocks but also about building the capacity to anticipate, prepare for, respond to, and recover from a wide range of challenges. The lessons learned from this crisis will likely shape how regional economies approach system-level human agency building in the face of future uncertainties, opening up new avenues for the research on regional economic resilience that has been much centered on the context of the 2008 financial crisis.

(E) *Relational perspective and resilience of chains*

In the context of deepening economic globalization and regionalization, China's contemporary economy, particularly the manufacturing sector, has become intricately embedded within global networks such as Global Production Networks (GPN) and Global Value Chains (GVC). As the international geopolitical and economic landscape becomes increasingly complex and challenging, global supply chains, industry chains, and value chains frequently encounter systemic shocks (Gereffi, 2020). In this deeply interconnected environment, China's economy also finds it progressively difficult to remain unaffected. To a large extent, the economies of OICs are not bounded into their own regional scale, but are inter-connected with other economies via cross-border industry organization , social networks, political nexus and policy relations. The economic resilience of regions is not an endogenous evolutionary process, but is continuously affected by all sorts of relations and networks at a broader sense (Hu et al., 2022). In this sense, faced with these realities, there is an urgent need for a shift in regional economic resilience research away from traditional evolutionary paradigms. Instead, there should be a renewed understanding of the resilience process and mechanisms from the perspectives of industy, value, innovation and supply chains. Here, we propose the idea of chain-based resilience which go against ideas from the conventional EEG approach. Table 3.3 shows the differences in the conceptualization between chain (also very much 'relational') resilience and evolutionary resilience.

3.6 Conclusions

In this paper we have made clear that ample research has been done on the restructuring OICs in China. Doubtlessly the restructuring of OICs in China is highly complex as it is not only affected by industrial-sectoral impact factors but also by political-institutional forces (Hu & Hassink, 2015). With the help of a critical review of Chinese-speaking articles in Chinese-speaking geography journals on this topic we have shown that the majority of papers focuses on policy-making recommendations related to the restructuring of OICs in China. On a critical note, we identified some clear deficiencies, such as the limited attention paid to dynamic multi-scalar interactions between firms, institutions, policies and places, as well as a strong focus on single cases. Most reviewed articles also disregard internationally recognized concepts and theories in the field, although much conceptual and theoretical work

Table 3.3 Conceptual comparison on chain resilience and evolutionary resilience

Item	Chain resilience	Evolutionary resilience
Disciplinary background	Development economics, macroeconomics, risk management, supply chain/value chain management, etc.	Economic Geography, Regional Economics
Research themes	Examining the adaptive, recuperative, and upgrading processes of economic chains and their constituents when facing disruptive risks	Exploring the evolutionary trajectory of regional economies as they navigate shocks, achieve recovery, and undergo growth amidst turbulent circumstances
Core perspective	Emphasis on multifaceted interrelationships	Focus on interfirm technological associations
Scale of object	Encompassing multiple levels (enterprises/institutions, industry sectors, regions, global context, etc.)	Centered at the regional level (cities, areas, countries, etc.)
Impact factor	External interconnections involving entities such as enterprises	External interconnections involving entities such as enterprises
Methodology	Primarily qualitative investigations	Emphasis on both qualitative and quantitative methodology

has been recently developed in English-speaking publications. On the basis of a review of these recently developed concepts and theories, we suggest some EEG concepts such as path dependence, lock-in, path creation and relatedness, as well as IEG, agency, geographical political economy, REG approaches as useful concepts for a better understanding of the processes and mechanisms of regional economic resilience of OICs in China.

Using these concepts as a framework for future research on the restructuring of OICs in China has in our view four main advantages. First, they help to overcome some of the identified deficits identified in this paper with the help of a critical literature review. Secondly, they will provide more in-depth and comprehensive explanations for the successes and failures of restructuring in different OICs in China. Thirdly, by providing more in-depth and comprehensive explanations researchers will also be able to derive more accurate and tailor-made policy recommendations going beyond the blueprint solutions produced in many existent studies. Fourthly, using and testing these international concepts in the Chinese context helps to make clear their explanatory boundaries. It will help to advance theoretical fundamentals by reflecting on underlying weaknesses of international theories, that is, 'theorizing back' at economic geography (Yeung & Lin, 2003; Yeung, 2007). Moreover, using them helps to build more solid intellectual bridges between Chinese and international geographers (Liu, 2009). Last but not least, taking the current COVID-19 crisis seriously and regarding it as an important window of contextual opportunity to reframe the existing conceptual framework of regional resilience. By giving more emphasis on contextual sensitivity and nature, regional economic resilience research needs to rethink the roles of structure and agency and their embedded contexts, and to re-conceptualize adaptation, adaptability and their interactive relationships in framing and shaping resilience.

Notes

1. According to the State Council, there are 118 resource-based cities widely distributed in China (see Fig. 3.1). These regions have strongly contributed to national economic growth by delivering energies and raw materials. Yet recently many of them have been confronted with the increasingly depletion of local natural resources. The national-state therefore initiated a set of supporting policies since 2008, in which 69 municipal regions were qualified as 'resource-exhausted cities' and supported with national financial aids for economic restructuring.
2. It has been used to articulate the over-loaded social responsibilities of SOEs under Mao's regime. SOEs were ordered to develop encompassing industrial complexes for self-contented local economies, characterized by the Chinese term, namely 'da er quan, xiao er quan' or 'chang ban da ji ti' known as 'collectively-owned factories under the control of SOEs'.
3. There is no strict definition of danwei, it generally refers to a place of employment in the context of pre-reform China, where employees were economically and socially constrained and even determined by particular institutions (Lu & Perry, 1997). Today, the danwei phenomenon is diminished but far from completely vanished, as SOEs still dominate particular sectors and economies.
4. Iron rice bowl refers to a permanent position in SOEs, authorities, and universities. This ideology can have a negative impact on self-employment and entrepreneurial activities (Zhang, 2000).

References

Arthur, B. (1994). *Increasing returns and path dependence in the economy*. The University of Michigan Press.

Bailey, D., Bellandi, M., Caloffi, A., et al. (2010). Place-renewing leadership: Trajectories of change for mature manufacturing regions in Europe. *Policy Studies, 31*(4), 457–474.

Birch, K., MacKinnon, D., & Cumbers, A. (2010). Old industrial regions in Europe: A comparative assessment of economic performance. *Regional Studies, 44*(1), 35–53.

Boschma, R., & Frenken, K. (2011). The emerging empirics of evolutionary economic geography. *Journal of Economic Geography, 11*(2), 295–307.

Boschma, R., & Lambooy, J. (1999). The prospects of an adjustment policy based on collective learning in old industrial regions. *GeoJournal, 49*(4), 391–399.

Boschma, R., Minondo, A., & Navarro, M. (2013). The emergence of new industries at the regional level in Spain. A proximity approach based on product-relatedness. *Economic Geography, 89*(1), 29–51.

Chen, X., Zhu, H. Y., & Zhang, L. H. (2006). Study on the industrial location questions and its target and mode of reconstruction in the Northeast old industrial base. *Economic Geography, 26*(3), 383–386. (in Chinese).

Cho, M., & Hassink, R. (2009). Limits to locking-out through restructuring: The textile industry in Daegu, South Korea. *Regional Studies, 43*(9), 1183–1198

Coe, N. M., Kelly, P. F., & Yeung, H. W. C. (2013). *Economic geography: A contemporary introduction* (pp. 510–512). Blackwell.

Cooke, P. (2011). Transition regions: Regional-national eco-innovation systems and strategies. *Progress in Planning, 67*(3), 105–146.

David, P. A. (1985). Clio and the economics of QWERTY. *American Economic Review, 75*(2), 332–337.

Dawley, S. (2014). Creating new paths? Offshore wind, policy activism, and peripheral region development. *Economic Geography*, 90(1), 91-112.

Ding, S.B., Sun, M. (2006). The development plight of Chinese resource-exhausted cities and the countermeasures of the central government. *Areal Research and Development*, 25(5), 1–5. (in Chinese)

Ding, S. B., & Sun, M. (2006). The development plight of Chinese resource-exhausted cities and the countermeasures of the central government. *Areal Research and Development, 25*(5), 1–5. (in Chinese).

Dong, L. J., & Zhang, P. Y. (2008). Industrial transformation and employment change of China's old industrial areas: A case study of Shenyang. *Scientia Geographica Snica, 28*(2), 162–168. (in Chinese).

Dunford, M., & Li, L. (2010). Chinese spatial inequalities and spatial policies. *Geography Compass, 4*(8), 1039–1054.

Economist. (2012). Reforming the north-east: rustbelt revival. *The Economist*, 2012–06–16.

Economist. (2015a). The north-east: back in the cold. *The Economist*, 2015–01–03.

Economist. (2015b). The cost of clean air. *The Economist*, 2015–02–07.

Fornahl, D., Hassink, R., Klaerding, C., et al. (2012). From the old path of shipbuilding onto the new path of offshore wind energy? The case of Northern Germany. *European Planning Studies, 20*(5), 835–855.

Frenken, K., Van Oort, F., & Verburg, T. (2007). Related variety, unrelated variety and regional economic growth. *Regional Studies, 41*(5), 685–697.

Garnaut, R., Song, L., & Yao, Y. (2006). Impact and significance of state-owned enterprise restructuring in China. *The China Journal, 55*(1), 35–63.

Gereffi, G. (2020). What does the COVID-19 pandemic teach us about global value chains? The case of medical supplies. *Journal of International Business Policy, 3*(3), 287–301.

Grabher, G. (1993). The weakness of strong ties: The lock-in of regional development in the Ruhr Area. In: Grabher G (Ed.), *The embedded firm: On the socioeconomics of inter firm relations.* (pp. 255–278). Routledge

Grillitsch, M., & Sotarauta, M. (2020). Trinity of change agency, regional development paths and opportunity spaces. *Progress in Human Geography, 44*(4), 704–723.

Han, J., Li, M. Z., & Huang, N. H. (2005). The framing mechanism of deindustrialization of old industrial bases in Northeast areas of China. *Economic Geography, 25*(1), 52–55. (in Chinese).

Han, J., Li, M.Z., Huang, N.H. (2005). The framing mechanism of deindustrialization of old industrial bases in Northeast areas of China. *Economic Geography*, 25(1), 52–55. (in Chinese)

Hassink, R. (2007). The strength of weak lock-ins: The renewal of the Westmünsterland textile industry. *Environment and Planning A, 39*(5), 1147–1165.

Hassink, R. (2010). Locked in decline? On the role of regional lock-ins in old industrial areas. In R. Boschma & R. Martin (Eds.), *Handbook of evolutionary economic geography* (pp. 450–468). Edward Elgar.

Hassink, R., Isaksen, A., & Trippl, M. (2019). Towards a comprehensive understanding of new regional industrial path development. *Regional Studies, 53*(11), 1636–1645.

Hassink, R., Klaerding, C., & Marques, P. (2014). Theoretical advancement in economic geography by engaged pluralism. *Regional Studies, 48*(7), 1295–1307.

Hassink, R., & Shin, D. H. (2005). The restructuring of old industrial areas in Europe and Asia. *Environment and Planning A, 37*(4), 571–580.

He, C., & Zhu, S. (2020). The principle of relatedness in China's regional industrial development. *Acta Geographic Sinica, 75*(12), 2684–2698. (In Chinese).

Hu, X. H. (2012). A review and prospect of the research on regional economic resilience. *Foreign Economies and Management, 34*(8), 64–72. (in Chinese).

Hu, X. H. (2014). State-led path creation in China's rustbelt: The case of Fuxin. *Regional Studies, Regional Science, 1*(1), 294–300.
Hu, X., & Hassink, R. (2017a). Exploring adaptation and adaptability in uneven economic resilience: A tale of two Chinese mining regions. *Cambridge Journal of Regions, Economy and Society, 10*(3), 527–541.
Hu, X., & Hassink, R. (2017b). Place leadership with Chinese characteristics? A case study of the Zaozhuang coal-mining region in transition. *Regional Studies, 51*(2), 224–234.
Hu, X., Li, L., & Dong, K. (2022). What matters for regional economic resilience amid COVID-19? Evidence from cities in Northeast China. *Cities, 120*, 103440.
Hu, X., & Yang, C. (2019). Institutional change and divergent economic resilience: Path development of two resource-depleted cities in China. *Urban Studies, 56*(16), 3466–3485.
Hu, X., Zhu, S., & Hassink, R. (2020). Beyond the "evolutionary approach": A critical review and paradigmatic reflections on the restructuring of old industrial areas. *Geographical Research, 39*(5), 1028–1044. (In Chinese).
Hu, X. H. (2015). From coal-mining to coal chemicals? Unpacking new path creation in an old industrial region in transitional China. *Growth and Change*, in press.
Hu, X. (2017). From coal mining to coal chemicals? Unpacking new path creation in an old industrial region of transitional China. *Growth and Change*. 48(2), 233-245.
Hu, X. H., & Hassink, R. (2015). Explaining differences in the adaptability of old industrial areas. In H. Ulrich (Ed.), *Routledge handbook of politics and technology*. Routledge.
Hudson, R. (2005). Rethinking change in old industrial regions: Reflecting on the experiences of North East England. *Environment and Planning A, 37*(4), 581–596.
Hudson, R., & Swanton, D. (2012). Global shifts in contemporary times: The changing trajectories of steel towns in China, Germany and the United Kingdom. *European Urban and Regional Studies, 19*(1), 6–19.
Li, X. J. (2009). New insights on China's regional economic research. *Economic Perspectives, 3*, 64–67. (In Chinese).
Li, X. J. (2022). History of the ideas on economic geography in China: Period focus, evolution and prospect. *Acta Geographic Sinica, 77*(8), 1873–1891. (In Chinese).
Li, G. C., & Li, Z. Q. (1996). Characteristics and causes of the 'Northeast Phenomenon.' *Economic Geography, 16*(1), 34–38. (In Chinese).
Lin, G. C. S. (2002). State policy and spatial restructuring in post-reform China, 1978–95. *International Journal of Urban and Regional Research, 23*(4), 670–696.
Liu, W. D. (2009). Where is the bridge? *Progress in Human Geography, 33*(1), 7–9.
Liu, W. D. (2020). The impacts of COVID-19 pandemic on the development of economic globalization. *Geographical Research, 39*(7), 1439–1449. (In Chinese).
Liu, W. D., Jin, F. J., Zhang, W. Z., et al. (2011). Progress in Chinese economic geography (2006–2011). *Progress in Geography, 30*(12), 1479–1487. (In Chinese).
Liu, W. D., & Lu, D. D. (2002). Rethinking the development of economic geography in mainland China. *Environment and Planning A, 34*(12), 2107–2126.
Liu, Z. G., & Dunford, M. (2012). Rejuvenating old industries in new contexts: The traditional Chinese medicine cluster in Tonghua, China. *Zeitschrift für Wirtschaftsgeographie, 56*(3), 185–202
Lu, X. P., & Perry, R. J. (1997). *Danwei: The changing chinese workplace in historical and comparative perspective*. Sharpe.
Lu, D. D., & Fan, J. (2010). *Regional development research in China: A roadmap to 2050* (pp. 39–60). Science Press.
Ma, L. J. C. (2002). Urban transformation in China, 1994–2000: a review and research agenda. *Environment and Planning A*, 34(9), 1545–1569
MacKinnon, D., Cumbers, A., Pike, A., et al. (2009). Evolution in economic geography: Institutions, political economy, and adaptation. *Economic Geography, 85*, 129–150.
MacNeil, S., & Steiner, M. (2010). Leadership of cluster policy: Lessons from the Austrian province of Styria. *Policy Studies, 31*(4), 441–455.

Maillat, D., Lechot, G., & Lecoq, B. et al. (1997). Comparative analysis of the structural development of milieux: The swach industry in the Swiss and French Jura Arc. In: R. Ratti, A. Bramanti, R. Gordon (Eds.), *The dynamics of innovation regions*. Ashgate.

Mao, H., Dong, J. G., Zhang, H., et al. (2008). Research on the layout of the old industrial base of Zhengzhou city and its coordinated deveopment. *Areal Research and Development, 27*(3), 36–38. (In Chinese).

Martin, R. (2010). Rethinking regional path dependence: Beyond lock-in to evolution. *Economic Geography, 86*(1), 1–27.

Martin, R. (2012). (Re)placing path dependence: A response to the debate. *International Journal of Urban and Regional Research, 36*(1), 179–192.

Martin, R., & Sunley, P. (2006). Path dependence and regional economic evolution. *Journal of Economic Geography, 6*(4), 395–437.

Moody, A. (2013). Digging deeper. *China Daily*, 2013–03–08.

Naughton, B. (1988). The third front: Defense industrialization in the Chinese interior. *The China Quarterly, 115*, 351–386.

NBSC (National Bureau of Statistics of China). (2014). *China Statistical Yearbook 2014*. Beijing: China Statistics Press. (in Chinese)

Neffke, F., Henning, M., & Boschma, R. (2011). How do regions diversify over time? Industry relatedness and the development of new growth paths in regions. *Economic Geography, 87*(3), 237–265.

OECD. (2015). *OECD economic surveys: China*. OECD Pubilishing.

Qiu, F., Yuan, H., Zhu, C., & Guo, M. (2018). The industrial transformation affects and influencing factors of regenerative resource-based cities in China. *Economic Geography, 38*(11), 68–77. (In Chinese).

Safford, S. (2009). *Why the garden club couldn't save youngstown: The transformation of the rust belt*. Harvard University Press.

Simmie, J., & Martin, R. (2010). The economic resilience of regions: Towards an evolutionary approach. *Cambridge Journal of Regions, Economy and Society, 3*(1), 27–43.

Sun, M., & Ding, S. B. (2005). An analysis of the institutional reasons on the recession of Chinese resource-based cities. *Economic Geography, 25*(2), 273–276. (in Chinese).

Tödtling, F., & Trippl, M. (2004). Like phoenix from the Ashes? The renewal of clusters in old industrial areas. *Urban Studies, 41*(5), 1175–1195.

Wang, M., Cheng, Z. M., Zhang, P. Y., et al. (2014). *Old industrial cities seeking new road of industrialization: Models of revitalizing Northeast China*. World Scientific.

Wang, L., Song, Z. Y., Yu, J. Y., et al. (2011). Strategic research on industrial transformation of resource-based cities: A case study on Kelamay. *Economic Geography, 31*(8), 1277–1282. (in Chinese).

Wang, K. Y., & Xiao, L. (2004). The culture problem on the economic transition of resource-exhausted cities in China. *Human Geography, 11*(1), 70–74. (in Chinese).

Wei, Y. D. (2000). Investment and regional development in post-Mao China. *GeoJournal*, 51, 169-179.

Xiao, J.S., Mao, F. (2008). A study on the evolutionary characteristics and developmental trends of resource-based cities in China. *Human Geography*, 100(2), 67-72. (in Chinese)

Yeung, H. W. C., & Lin, G. C. (2003). Theorizing economic geographies of Asia. *Economic Geography*, 79(2), 107-128.

Yeung, H. W. C. (2007). Remaking economic geography: insights from East Asia. *Economic Geography*, 83(4), 339-348.

Yin, J., Luo X.L., Chen Y.Q., et al. (2010). Entrepreneurial city theory and the transition of mineral industrial city: a case study of Ma'anshan City. *Scientia Geographica Sinica*, 30(3), 329–335. (in Chinese)

Zhang, P.Y. (2004). New industrialization and policies for the transformation of the Northeast China's old industrial base. *Economic Geography*, 24(6), 784–787. (in Chinese)

Zhang, P.Y. (2006). Institutional and cultural factors in affecting the restructuring of Tiexi industrial district in Shanyang, China. *Human Geography*, 88(2), 45–49. (in Chinese)

Zhang, P.Y. (2008). Revitalizing old industrial base of Northeast China: Process policy and challenge. *Chinese Geographical Science*,18(2), 109-118.

Zhu, H., Liu Y.Y., Bao J.G. (2002). The exploration of old industrial cities′ tourism planning: a case study of Huangshi City in Hubei Province. *Economic Geography*, 22(2), 252–257. (in Chinese)

Zhang, W.Z., Wang, D., Yu J.H. (2011). A study on the development path and pattern of continued and alternative industries in resource-based cities. *Bulletin of Chinese Academy of Sciences*, 26(2), 134–141. (in Chinese)

Zhang, W., Yu, J., & Li, J. (2016). Driving factors and mechanism of resource-exhausted city transformation. *Bulletin of Chinese Academy of Sciences, 31*(1), 92–100. (In Chinese).

Zhang, J., & Peck, J. (2016). Variegated capitalism, Chinese style: Regional models, multi-scalar constructions. *Regional Studies*, 50(1), 52-78.

Zhou, Y., & He, C. (2019). Evolution of urban export product in China. *Acta Geographic Sinica, 74*(6), 1097–1111. (In Chinese).

Zhang, W., & Yu, J. (2023). Policy evolution and transformation effect analysis of sustainable development of resource-based cities in China. *Journal of Natural Resources, 38*(1), 22–38. (In Chinese).

Chapter 4
Industrial Structure or Agency: What Affects Regional Economic Resilience Under the 2008 Economic Crisis? Evidence from Resource-Based Cities in China

4.1 Introduction

Although regional resilience has been criticized as a fuzzy concept, lacking a solid definition and meaning (Hassink, 2010; Martin, 2018), it has generated great scholarly interest in economic geography and regional studies. This has been especially the case after the 2008 global financial crisis (Cainelli et al., 2019; Fröhlich & Hassink, 2018). The diverse adaptation processes and adaptability of regional economies to crises have drawn much attention to the determinants of economic resilience (Hu & Hassink, 2020; Martin, 2012; Martin & Sunley, 2020). Several key factors in shaping the process and ability of regional resilience have been identified, including economic or industrial structure, human capital, innovation, quality of government, institutional environment etc. (Brown & Greenbaum, 2017; Ezcurra & Rios, 2019; Martin et al., 2016). Industrial structure, in particular, as one of the most fundamental factors, has renewed the debates on whether specialization or diversification is "constraining or enabling" regional development in essence in an uncertain context (Martin et al., 2016).

Preoccupied by the firm-centric and endogenously-based conceptualizations of the evolutionary approach, the existing account on the topic has led to two key interrelated assumptions: first, regional structural factors, as historically inherited endowments and locally anchored assets, determinate a region's shock-absorptive nature (resistance ability) and recoverability—as two key dimensions of regional resilience (Martin & Sunley, 2015). Second and moreover, resilient regions are regions with structural advantages where diversified industries, thick institutions, high social and technological relatedness, innovative/learning environments are present (Balland et al., 2015; Xiao et al., 2018).

(This chapter is a reprint version from: Tan, J., Hu, X., Hassink, R., & Ni, J. (2020). Industrial structure or agency: What affects regional economic resilience? Evidence from resource-based cities in China. Cities, 106, 102906.)

X. Hu, *Uneven Economic Resilience of Old Industrial Cities in China*, Urban Sustainability, https://doi.org/10.1007/978-981-99-9279-9_4

One key pitfall, as many scholars have increasingly criticized, is that the evolutionary conception of resilience reflects a superficial "regional fetishism", viewing regions as structured entities while overlooking the dynamic role of extra-local factors in shaping resilience (Evenhuis, 2017). Moreover, increasing evidence suggests that structurally weak regions (e.g., peripheral regions, single industry towns) are not doomed to be weak in resilience. Weak structure may stimulate agents to work in broader contexts and generate collective agency for higher resilience (Bristow & Healy, 2014; Hu & Yang, 2019). This is particularly the case for structurally weak resource-based cities (RBCs hereafter) in China. Although RBCs' industrial structure is highly rigid and singular, and embedded in less marketized and low-tech based institutional contexts, existing, mainly qualitative resilience research on Chinese RBCs has a stronger focus on the enabling role of state agency, leaving the role of industrial structure underscored (Hu, 2017; Hu & Hassink, 2017a; Hu & Yang, 2019).

Thus, the question on what determinates regional resilience is unclear and highly contested, particularly for structurally weak regions. This article aims to explore the role of industrial structure vs. agency in affecting regional economic resilience with empirical evidence from Chinese RBCs. We measure the economic resilience of RBCs after 2008, by decomposing resilience into an industrial structure effect and an agency effect (which explains how much of the change in resilience is due to regional non-structural forces). By using a panel regression model, we examine the impact of industrial structure in terms of diversification and specialization in different stages of the crisis. The results address three key questions: (1) do regions with a typical specialized economy suffer more strongly during a recession? (2) how does industrial structure affect the resilience of RBCs? and (3) what kind of industrial structure enables resilience? The remainder of the paper is organized as follows. Section 4.2 reviews the literature on regional resilience with a particular focus on its conceptualizations and measurements. Section 4.3 offers a conceptual framework for understanding the determinants of regional resilience and proposes key hypotheses for empirical analysis. Section 4.4 introduces the research area and methodology. After discussing empirically how diversification and specialization affects resilience respectively in Sects. 4.5, 4.6 and 4.7 concludes.

4.2 Regional Economic Resilience and Its Measurements

The word resilience comes from the Latin 'resilire', which means to rebound or to leap back (Martin, 2012). It originates from ecology (Holling, 1973), and then traveled to engineering, psychology, more recently, to social sciences (Faggian et al., 2018). Resilience is more than a metaphor but less than a solid theory, which nevertheless helps us to (re)think regional economies in a dynamic, holistic and systematic way (Hassink, 2010; Hu & Hassink, 2020). In general, resilience includes three key types: engineering resilience, ecological resilience and adaptive resilience (Martin, 2012). Engineering resilience refers to the ability of a system to return its assumed stable equilibrium, which can be measured by the return speed to equilibrium. Ecological resilience is defined as the extent or scale of the disturbance that the system can absorb before it is de-stabilized and move to another stable equilibrium (Holling, 1973). Adaptive resilience means the ability of a system to undergo anticipatory or reactionary reorganization of a system's structure to minimize the impact of a shock (Martin, 2012). In short, the three types of resilience refer to the ability of bouncing back from shocks, absorbing shocks and positive adaptation in response to shocks, respectively (Martin & Sunley, 2015). The former two types adopt an equilibrium-based approach, which are criticized by economic geographers (Balland et al., 2015). The latter is conceptualized as a nonequilibrium and adaptation process of constant continuity and change, characterized by open-ended adaptability of developing different outcomes and trajectories (Hu & Hassink, 2017a, 2020). Such a process involves four main inter-related dimensions, including resistance, recovery, re-orientation and renewal (Martin, 2012). Martin and Sunley (2015) further extended the resilience definition with normative meaning by introducing the notions of vulnerability, resistance, robustness and recoverability. Vulnerability means the sensitivity or propensity of a region's economy related to pre-shock regional growth structure. Resistance refers to the depth of direct response to a shock, which is not only related to the nature of the shock but also to the attribute of regional structural factors (Simmie & Martin, 2010). Robustness and recoverability stand for adaptive agency related the actions and strategies to shocks and recovery from the downturn (Martin & Sunley, 2020).

Due to its non-equilibrium nature, regional economic resilience's measurement is challenging. It varies in method, scale and unit of analysis (Courvisanos et al., 2016; Sensier et al., 2016). First, most measurement studies have mainly focused on the assessment of "recoverability" from the 2008 global crisis. Different resilience indicators are used such as (un)employment, GDP, exports records, and other composite indictors (Di Caro, 2015; Doran & Fingleton, 2016; Fingleton et al., 2012). Nevertheless, these studies all reveal a highly heterogeneous spatial pattern among regions and countries. Second, recent work has been paid much attention to explaining why some regions are more resilient than others. Giannakis and Bruggeman (2017b), for instance, found that rural areas had higher resilience in Greece, because rural areas, and perhaps, other economically self-contained places, are less likely to be hit by the 2008 global crisis, compared to globalized, manufacturing regions. Third, most measurement research is based in a European context, whereas related work in emerging economies is scarce. In a handful of resilience research in China, less-favored RBCs have been the core study object (Guan et al., 2018; Hu, 2017; Hu & Hassink, 2017a; Hu & Yang, 2019). In contrast to what mainstream research argues, these studies move beyond the structure-centered view, but rather, highlight the positive role of agency in determining regional resilience.

Economic resilience of regions in different national contexts shows a heterogeneous pattern. A complex set of impact factors, which can be economical, social, institutional, and in structural or agentic forms, from local and extra-local scales, shapes the nature of regional resilience (Martin & Sunley, 2020). However, the answer to what determines resilience remains unclear (Hu & Hassink, 2020).

4.3 Determinants of Regional Resilience: A Conceptual Framework

(1) *Industrial structure matters: diversification matters*

As has become clear in the previous section, a complex array of factors affects regional economic resilience. Most of the literature agrees that economic structure strongly matters for resilience, since it makes a regional economy directly exposed to external shocks (Martin et al., 2016). Economic structure represents a region's economic profile, constructed by regionally inherent and historically inherited elements. It can be decomposed into several inter-related factors, such as economic openness, export scale/level, industrial structure, productivity, technological level, economic linkages, and policy regime (Martin & Sunley, 2015). Industrial structure has been regarded as the most fundamental factor for resilience. The question of whether specialization (or diversification) is good or not for regional resilience is the most debated yet unsettled topic in the literature.

Many studies highlight that it is rather diversification making anti-risk ability against shocks of any kind stronger (Xiao et al., 2018). Christopherson et al. (2010) proposed that a diversified economic base, not over-reliant on a single industry enabled a region to adjust and adapt, since different industries had different responds to the recession, and could spread the risks to distributed activities. Similarly, Brown and Greenbaum (2017) examined the effect of industrial diversification on economic resilience in Ohio counties in the U.S. and concluded that counties with higher industrial diversification experienced higher resilience to external shocks, while specialized counties were less resilient due to their lack of ability to absorb shocks. Di Caro (2015) also discovered the positive impact of diversification over specialization on resilience in Italian regions. Moreover, it is evident that a higher level of related diversification is likely to create new industrial paths with a higher resilience (Balland et al., 2015; Boschma, 2015). Unrelated diversification reflects the virtue of "not putting all eggs in the same basket" (Kemeny & Storper, 2015) and shows more ability of shock absorption than specialization (Zhu et al., 2017).

(2) *Agency also matters: towards a conceptual framework*

Recently, regional resilience research has shown that non-industrial structural factors can play a key role in shaping regional resilience (Martin & Sunley, 2020). A common understanding is that place-specific actors including policy arrangements, institutional settings, knowledge bases and social capital can generate their independent impact on resilience (Bristow & Healy, 2018; Kitsos et al., 2019; Martin & Gardiner, 2019; Sedita et al., 2017). Moreover, the role of extra-local factors, particularly related to national policy, is increasingly taken into account in the research (Eraydln, 2016; Giannakis & Bruggeman, 2017b; Webber et al., 2018).

In fact, a growing body of research on regional resilience has appeared to shift from the traditional "structure-focused" view to an "agency" perspective (Boschma, 2015; Bristow & Healy, 2020). Less-favored regions lacking structural endowments and advantages, therefore, have become "fertile grounds" for the re-theorizing of regional resilience. In contrast to the mainstream idea of "industrial structure matters", this research body provides a set of novel evidence that structurally weak less-favored regions can also be resilient. While the role of industrial structure seems to be neutralized or even overlooked, resilience is determined by human agency. For instance, the emergence of offshore wind power industry and the subsequent industrial restructuring in some regions of Northern Germany and Northeast England is attributed to local political leaders who lobby for broader support (Dawley, 2014; Fornahl et al., 2012). Martin and Martin (2017) found that policy-makers and their actions can overcome the deficits related to the single-industry structure in two industry towns of Sweden and generate stronger resilience. Other non-firm actors and their agencies, such as social entrepreneurs, community leaders and labor union actors, have been particularly highlighted as key sources and drivers of regional resilience (Evenhuis, 2016; Fredin et al., 2019; Hervas-Oliver et al., 2011; Safford, 2009).

Only a few studies think both industrial structure and agency are important determinants of regional resilience. Bristow and Healy (2020) suggest that economic resilience as a complex process is driven by a combination of structural and agential factors. Martin et al. (2016) suggest that in the long term, industrial structure mainly influences resistance and recoverability, whereas agency is likely to have more impact on renewal and restructuring. Hu and Yang (2019) argue that resilience as a process is conceptualized by adaptation and adaptability. Adaptation refers to agency maintaining existing industrial structure while adaptability means agency building ability for structural change, both of them can shape resilience.

To be clear, regional economic resilience can be generally affected by two main types of effects: the industrial structure effect and the agency effect (Bristow & Healy, 2020; Hu & Hassink, 2020; Martin & Sunley, 2015). To better understand the determinants of regional resilience, we propose a conceptual framework that takes both industrial structure and agency effect seriously. The framework can be understood by three key building blocks: Balland et al., (2015) the industrial structure effect is regionally inherent and inherited, generated by structural and supporting elements such as policy regime, technological profile, business modes, industrial legacies, labor profile, financial institutions and entrepreneurship culture (Boschma, 2015; Martin & Sunley, 2015). The agency effect is made by purposive actions of actors and is shaped by their motives and capability of making changes and impacts on structural elements. Here, agency mainly involves three types across a wide range of organizations and institutions (state, quasi-state, and firm), namely, place leadership, institutional entrepreneurship and firm-based innovative entrepreneurship (Breathnach et al., 2015; Grillitsch & Sotarauta, 2019). It is important to highlight that regional resilience is a process involving the interplay between agency and industrial structure in a co-evolutionary manner (Gong & Hassink, 2019). Resilience, in other words, evolves over time and space in which agency and industrial structure may play different roles.

(3) *The resilience of RBCs in China: two key hypotheses*

RBCs are cities whose economies depend primarily on exploiting and processing natural resources, and since the non-renewable character of these natural resources, the research on RBCs mainly concentrated on sustainable development, urban transformation, and economic revitalization (Li et al., 2015; Zhang et al., 2011). RBCs have encountered more recession due to single and specialized industrial structure and thin institutions (The Economist, 2014). It is documented that the GDP growth rate of China dropped by 8% points from 2007 to 2016, of which 54 out of 114 prefecture-level RBCs dropped more than 8% points. Even worse, some RBCs dropped more than 20% points, others experienced severe negative growth, such as Fuxin, Anshan, Tonghua and Benxi (Hu & Hassink, 2017a). Thus, studying the resilience of RBCs has been a topical theme with both scholarly and policy values in China.

Most research on economic resilience of RBCs in China is based on a qualitative approach highlighting the role of state agency. Hu and Hassink (2017b) proposed a conceptual framework of long-term regional economic resilience by developing a typology of the relationships between adaptation (resilience processes) and adaptability (resilience capability). They applied this typology to two Chinese RBCs, namely Zaozhuang and Fuxin, and found that Zaozhuang with related diversification was more resilient than Fuxin with unrelated diversification. Hu and Yang (2019) put forward an analytical framework that addresses how different patterns of institutional change in regions lead to uneven economic resilience. Only a few scholars adopt quantitative methods to study resilience in China. Tan et al. (2017), for instance, measured the economic resilience of RBCs in Northeast China in terms of resistance and recoverability and explored their determinants. They found that Chinese RBCs presented poor resistance, and their industrial structures had a weak and ambiguous effect on economic resilience.

To sum up, the existing research is very much case-based, highlighting the role of state agency and institutions, while lacking a structure-sensitive perspective on Chinese RBCs. More importantly, Previous research usually involves a mix set of industrial structures and types at the national or local level, however, few studies consider a specific type of cities, such as RBCs, where industrial structure is often more rely on a certain type industry, such as mining and mineral processing industry. The question of how and to what extent the industrial structure affects regional economic resilience in RBCs is still unclear. Moreover, given the fact that the existing knowledge of quantitative resilience research is mainly derived from industrially diversified regions in market-led developed contexts, exploring the effect of industrial structure on resilience of RBCs in China thus may generate some new insights to the literature.

Based on our literature review and conceptual framework, we recognize that regional industrial structures play different roles in different regions and at different stages of a crisis. In other words, RBCs do have a relatively low degree of industrial diversification compared to other types of regions (e.g., metropolitan areas). However, within RBCs, the degree of diversification or specialization varies from city to city. Considering the industrial structure characteristics of and their differences among RBCs, we propose two hypotheses: First, RBCs with their disadvantaged industrial structure should encounter more recession in the crisis than other cities. Second, within the RBCs, cities with higher industrial diversification experience higher resilience, while regions with higher specialization have lower resilience. The two hypotheses will be tested in the following Sect. 4.3. Materials and Methods.

4.4 Materials and Methods

4.4.1 Study Area and Data

According to the "Plan of Sustainable Development for Resource-based Cities in China (2013–2020)" issued by the State Council, there are 262 RBCs in China, including 126 prefecture-level cities, 62 county-level cities, 58 counties, and 16 municipal districts. Given data availability, 114 RBCs are selected as study area (Fig. 4.1). In 2016, the total GDP of 114 RBCs was 16,734.7 billion, accounting for 22.49% of the national total. Their average GDP per capita was 40,536 yuan, much lower than the national average (53,980 yuan). The ratio of the three sectors output in RBCs was 11.2: 48.2: 40.6 in 2016, which has a higher proportion in the primary and secondary sector than the national average. Our data was derived from China City Statistical Yearbook and China Statistical Yearbook from 2008 to 2017. Urban vector data was obtained from the Institute of Earth System Science's data sharing platform (http://www.geodata.cn).

4.4.2 Methods

(1) *Measuring economic resilience*

In this article, we follow the method proposed by Martin et al. (2016) to measure economic resilience in terms of resistance using gross domestic product (GDP) data. Since resilience is interested in how different regions are affected by a common recession (nationwide), then the expected economic change in region r during a recessionary period, would be given as:

$$(\Delta \mathrm{E}_{\mathrm{r}}^{\mathrm{t+k}})^{\mathrm{expect}} = \sum_{i} E_{ir}^{t} * g_{N}^{t+k} \tag{4.1}$$

where $(\Delta \mathrm{E}_{\mathrm{r}}^{\mathrm{t+k}})^{\mathrm{expect}}$ is the expected recession value of national output from time t to t + k, and g_N^{t+k} is the GDP growth rate in recession period of nationwide. E_{ir}^{t} is the output value of industry i in region r at starting time t. Then the measure of regional

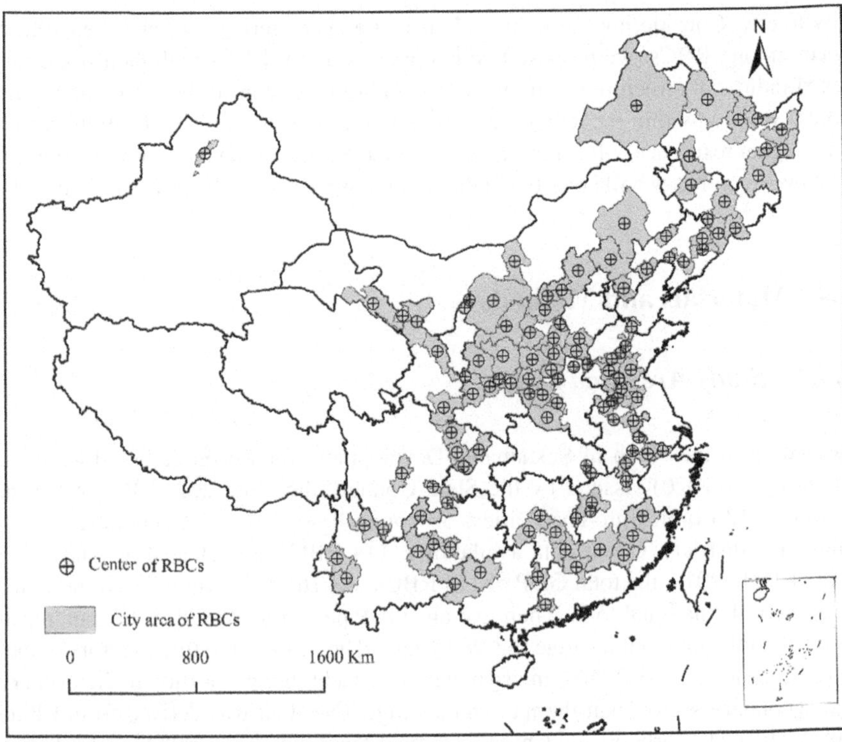

Fig. 4.1 The location of RCBs in China (authors' own composition)

resistance can be expressed as:

$$\text{Resis}_r = \frac{(\Delta E_r^{t+k}) - (\Delta E_r^{t+k})^{expect}}{\left|(\Delta E_r^{t+k})^{expect}\right|} \tag{4.2}$$

where ΔE_r^{t+k} is the recession value of region r from time t to t + k. By definition, the value of Resis_r was around zero, and when the value of Resis_r was positive, a region is more resistant to recession than the national average level. This study focuses on the global financial crisis in 2008. The GDP growth rate in RBCs experienced roughly the same development trend as China, which peaked in 2007; declined in 2008, recovered briefly in 2010; and declined again; and after 2012 was stable. Since the brief recovery in 2010 was mainly caused by a 4 trillion CNY investment plan, which was proposed by the central government to prevent rapid economic recession. Therefore, we define the period from 2007 to 2016 as recession period and measure the economic resistance to indicate the resilience.

(2) *Decomposing economic resilience*

The shift-share analysis method (SSM), usually employed in regional science and urban studies, decomposes employment or GDP growth (or decline) in a region/city over a given time period into three components: national growth effect, industry mix effect and competitive effect. SSM helps to answer why employment or GDP is growing or declining in a local industry, cluster, or occupation. The traditional SSM formula can be written as:

$$\Delta \mathrm{E}_{\mathrm{r}}^{\mathrm{t+k}} = NS_r + IM_r + RS_r \tag{4.3}$$

$$\begin{aligned} NS_r &= \sum\nolimits_i E_{ir}^t * g_N^{t+k};\ IM_r = \sum\nolimits_i E_{ir}^t \left(g_{iN}^{t+k} - g_N^{t+k}\right) \\ RS_r &= \sum\nolimits_i E_{ir}^t \left(g_{ir}^{t+k} - g_{iN}^{t+k}\right) \end{aligned} \tag{4.4}$$

where NS_r, IM_r and RS_r is national growth effect, industry mix effect and competitive effect, respectively. E_{ir}^t and g_N^{t+k} have been explained before, and g_{iN}^{t+k} is the growth rate of industry i, nationwide. g_{ir}^{t+k} is growth rate of industry i in region r. NS_r, the national effect or component, means the change of regional GDP in period t to t + k, that would be expected if output in each of the region's industries change at the same rate as the national rate of output change. IM_r is the portion of the GDP change attributed to the performance of the specific industry at the national level. RS_r explains how much of the change is due to some unique competitive advantage that the region has. We use SSM to decompose regional economic resilience into two parts as follows. Both sides of the Eq. (4.3) minus

$$\sum\nolimits_i E_{ir}^t * g_N^{t+k}.$$

$$\begin{aligned} \Delta Ert = k - \sum\nolimits_i E_{ir}^t * g_N^{t+k} &= \sum\nolimits_i E_{ir}^t \left(g_{iN}^{t+k} - g_N^{t+k}\right) \\ &+ \sum\nolimits_i E_{ir}^t \left(g_{ir}^{t+k} - g_{iN}^{t+k}\right) \end{aligned} \tag{4.5}$$

Then both sides of the equation divided by $\left|\sum_i E_{ir}^t * g_N^{t+k}\right|$ as follows.

$$\frac{\Delta \mathrm{E}_{\mathrm{r}}^{\mathrm{t+k}} - \sum_i E_{ir}^t * g_N^{t+k}}{\left|\sum_i E_{ir}^t * g_N^{t+k}\right|} = \frac{\sum_i E_{ir}^t \left(g_{iN}^{t+k} - g_N^{t+k}\right)}{\left|\sum_i E_{ir}^t * g_N^{t+k}\right|} + \frac{\sum_i E_{ir}^t \left(g_{ir}^{t+k} - g_{iN}^{t+k}\right)}{\left|\sum_i E_{ir}^t * g_N^{t+k}\right|} \tag{4.6}$$

where $\frac{\Delta \mathrm{E}_{\mathrm{r}}^{\mathrm{t+k}} - \sum_i E_{ir}^t * g_N^{t+k}}{\left|\sum_i E_{ir}^t * g_N^{t+k}\right|}$ just is the formula of resilience show in Eq. (4.2).

Then the economic resilience in terms of economic resistance or recoverability is decomposed into an industrial structure effect (IS) and a regional competitive effect (RC). Industrial structure effect means the change of GDP in a recession (or recovery) allowing for the fact that some national industries may change at faster rates than national GDP as a whole. It thus captures that part of a region's relative resistance to and recoverability from a recession, that is attributable to the region's mix of more and less nationally resilient industries. The more a region's economic structure is dominated by sectors that nationally are more resilient to recessions, the more resilient the region will be to a recession compared to the national economy. The regional competitive effect captures differences in behavior between the industries in a region and those same industries nationally.

(3) *Measuring diversification and specialization of industrial structure*

We apply several methods to quantify industrial structure from diversification and specialization by using the employment data of 19 sectors. This would help to check the robustness of results with respect to a variation in measurement. Industrial diversification is measured via the calculation of a diversity index (DIV), and the index value is given by Eq. (4.7):

$$\mathrm{DIV} = \sum_{i=1}^{n}\left(\frac{e_{ri}}{e_r}\right) * \ln\left(\frac{1}{e_{ri}/e_r}\right) = -\sum_{i=1}^{n}\left(\frac{e_{ri}}{e_r}\right) * \ln\left(\frac{e_{ri}}{e_r}\right) \tag{4.7}$$

where DIV is the industrial diversification in region r, and n is the number of total industries; e_{ri} is the employment of industry i in region r and, and e_r is the total employment in region r; ln means the natural logarithm. The higher the index value, the greater the industrial diversification. If the index value is zero, it means the region has only one industry.

Location quotients (LQ) are used to measure the degree of specialization of industrial structures. First, we try to examine how a specialized local industry affects economic resilience. We attempt to find whether certain industries can increase the resilience probability of RBCs. Then the LQ is calculated, and as seen in Eq. (4.8):

$$\mathrm{LQ}_{ri} = \frac{e_{ri}/e_r}{e_{Ni}/e_N} \tag{4.8}$$

where e_{ri} is employment of industry i in region r, and e_r is total employment in region r; e_{Ni} is the employment of industry i in the state, and e_N is total employment in the state. LQ_{ri} is the location quotient of industry i in region r. If the LQ_{ri} is larger than one, it means industry i in this region is specialized. In this article, we select seven industries to analyze the effects of the specialization of a certain industry on resilience, including Mining, Manufacturing, Construction, Wholesale and Retail, Financial Intermediation, Education, Technical Service and Geologic Prospecting. Because all these industries account for high proportions of employment and play an important role in the development of RBCs. Then to examine the industry specialization of the region that not only focuses on a certain industry, we propose the regional LQ as follows.

$$LQ_r = \sum_{i=1}^{N} LQ_{ri}^2 \tag{4.9}$$

where the LQ_r means the industrial specialization of region r. The higher the index value, indicating the region is more specialized in certain industries. Finally, we used LQ_r and LQ_{ri} to represent the industrial specialization.

(4) *The regression models*

Since the spatial distribution of RBCs is discontinuous, this article does not consider spatial correlation characteristics. We employ a panel regression analysis to detect the effect of industrial structure in terms of diversity and specialization on economic resilience. Regional economic resilience is determined by a complex array of factors. Considering the availability of data, we set five control variables based on the determinant analysis framework proposed by Martin and Sunley (2015) and other related research (Brown & Greenbaum, 2017; Cainelli et al., 2019; Giannakis & Bruggeman, 2017a). We choose per capital GDP (PCGDP) to capture the level of economic development, and choose the population density (Pop_den) to capture the distribution of population and it can also reflect the human resource. We also use unemployment rate (Unemployment) to capture the labor condition. We select the ratio of foreign direct investment to GDP (FDI) to reflect the financial arrangement. Finally, the governance usually measured by the quality of government and others in European (Ezcurra & Rios, 2019), however, in China, no official data on government quality is available. Considering the proportion of fixed asset investment was very high (77.5% in 2017) in China, and was dominated by the government, the ratio of fixed asset investment to GDP (Fix_Invest) was selected to express the ability of governance. Then the model of panel regression analysis is as follows.

$$\begin{aligned} y_{it} = \alpha_i + \beta_0 + \beta_1 \text{industry}_{it} + \beta_2 \text{PCGDP}_{it} + \beta_3 \text{Pop\_den}_{it} + \beta_4 \text{FDI}_{it} \\ + \beta_5 \text{Fix\_Invest}_{it} + \beta_6 \text{Unemployment}_{it} + u_{it} \end{aligned} \tag{4.10}$$

where y_{it} is dependent variable (the economic resilience in terms of resistance in region i of t year). industry_{it} is the core independent variable of industrial structure, including DIV, LQ_r and seven industries' LQ_{ri} and then PCGDP_{it} ...... Unemployment_{it} are five control variables. β_0, β_1 ...... β_6 are coefficients, α_i is non-observable variable, and u_{it} the error, varies over region i and time t. the error u_{it} is used to decide to select fixed effects or random effects model.

4.5 Main Results

4.5.1 *Characteristic of Economic Resilience in RBCs*

Based on the formulas (4.1) and (4.2), we calculated the economic resistance of 114 RBCs from 2008 to 2016. The results are shown in Fig. 4.2 to represent the economic resilience. Based on the value of economic resistance, RBCs were divided into green and red categories. The green category indicated values higher than the national average while the red indicated values lower than it. We find that at the beginning of the economic crisis, the economic resilience of RBCs was generally high, with only 27 cities having lower-than-average economic resistance. The distribution of low economic resistance was random. In 2016, the number of cities with low economic resilience increased to 52 which are mainly based in Northern China, including Liaoning, Jilin, Heilongjiang, Shandong, Hebei, Henan, Shanxi and Ningxia. We thus may conclude that the effect of economic crisis on RBCs differed during different stages of the crisis, and the RBCs with a structurally weak and specialized economy not always confronted serious recession.

To explore the different reaction of RBCs to the Great Recession at different stages, we calculated the mean of economic resilience and the number of RBCs with low value from 2008 to 2016 (Fig. 4.3). The mean value of economic resilience of RBCs was high in general, and the number of cities with lower-than-average economic resistance was limited, even in the lowest year 2016, still more than half of RBCs having higher economic resilience. This was mainly because most RBCs had higher GDP growth rates than the national average level at the beginning of crisis, and the growth value of regional output was higher than nationally expected. The mean of economic resilience showed an inverted U-shaped development trend, which increased from 2008 to 2012, and then declined after 2012. This indicates the global financial crisis has not brought severe shocks to RBCs directly, and the RBCs encounter deep recession in the post-financial crisis era. At the beginning

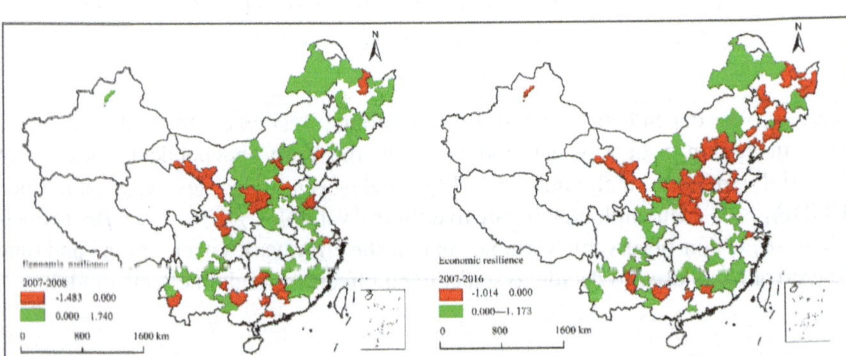

Fig. 4.2 The spatial distribution of economic resilience RBCs to the global financial crisis

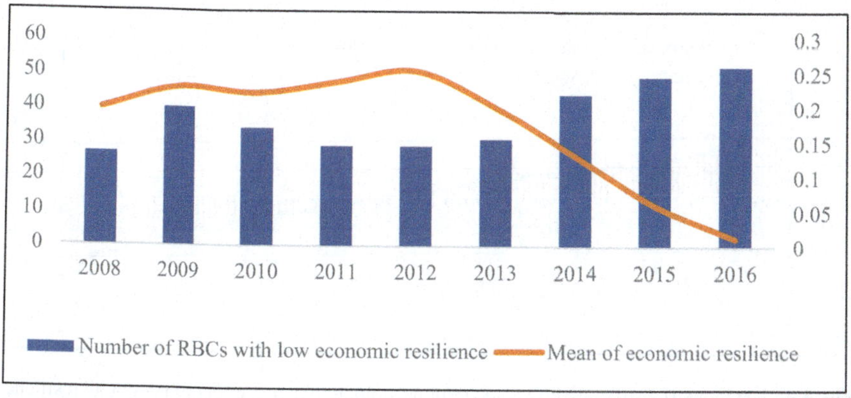

Fig. 4.3. The development of economic resilience from 2008 to 2016.

of the crisis, the high economic resilience of most RBCs can be attributed to their industrial structure. Indeed, the impact of the financial crisis on China was mainly on the real economy, because of the declining exports, especially of manufactured goods. However, RBCs are more dependent on primary goods, and the export destinations are mainly concentrated in coastal regions. Besides, such phenomenon can be also attributed to the remote location and the 4 trillion Investment Plan. After 2012, the GDP growth rate of RBCs dropped rapidly, and the mean of economic resilience began to decline and closed to zero in 2016. In short, the RBCs in China with typical specialized economy generally had high economic resilience, especially at the beginning of the global financial crisis, and the recession in RBCs was delayed.

4.5.2 Decomposing the Economic Resilience of RBCs

Based on the formula (4.6), the economic resilience of 114 RBCs was decomposed into industrial structure effect and regional competitive effect using the three industries data (in Fig. 3.1). The RBCs were ranked from left to right by the value of economic resilience. Several features are evident from Fig. 3.1. First, the regional competitive effect has a similar development trend with economic resilience and the absolute value of the regional competitive effect is higher than that of the industrial structure effect. This indicates that the economic resilience is mainly determined by the regional competitive effect in RBCs. Secondly, most cities' industrial structure effect is negative and the difference between different cities is not big. Only three cities, namely Lijiang, Baotou and Mudanjiang, have positive industrial structure effect, and this is mainly because only these cities had a higher proportion of the tertiary sector than the national average level in 2007. The effect of the industrial structure in RBCs has been what could be expected, namely that the effect has been negative in RBCs being more dependent on the secondary sector, especially mining,

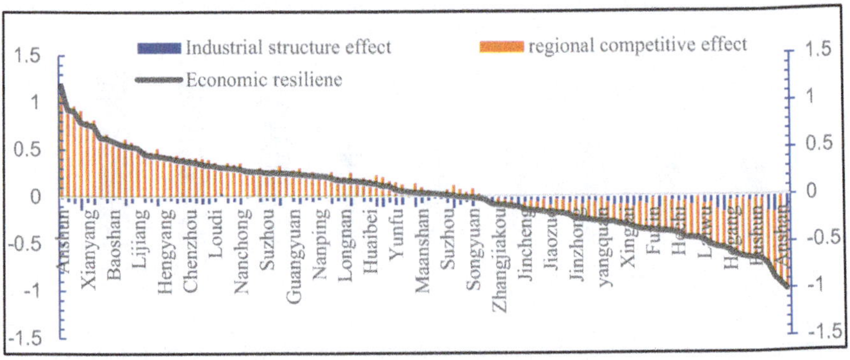

Fig. 4.4 The industrial structure effect and regional competitiveness effect of economic resilience

and very few RBCs benefitted from a favorable industry mix effect. Third, more than half of RBCs, in total 67, had a positive regional competitive effect, which indicates local characteristics, such as abundant resources, remote location, and central policy support helped the RBCs to be more resistant to the global financial crisis.

Overall, the findings in Fig. 3.1 suggest that, generally, the industrial structure had a negative effect on regional economic resilience in RBCs, and that the regional competitive effect seems to have played a dominant role in shaping how RBCs react to the economic crisis. The industrial structure effect mainly focuses on the initial state of RBCs' industrial structure and the change of national industrial structure, while the regional competitive effect focuses on the change of industrial structure both on nationwide and RBCs. Furthermore, in this article we decomposed the economic resilience only into three industries. As Martin (2016, p. 573) stated "A finer level of disaggregation might reveal a larger role for industrial mix". Thus, we further analyze the regional competitive effect of different industries (Fig. 4.4).

Based on formula (4.6), we further analyze the industrial structure effect and regional competitive effect of different industries. The results show that the mean values of the first, second and tertiary sectors effect is −0.041, −0.123 and 0.104 respectively, which means that the tertiary sector has a positive effect, while the first and second sector have negative effects on the industrial structure effect. However, the development of the tertiary sector in RBCs is very backward. Therefore, the industrial structure effect in RBCs is in general negative. Furthermore, several interesting features can be found in Fig. 4.5, where the regional competitive effect of different industries is shown. First, the regional competitive effect of RBCs was mainly determined by the secondary sector, the mean of the absolute value of the three sectors is 0.034, 0.255 and 0.109. Second, the Std. Deviation of the second sector regional competitive effect is highest (0.304), followed by the tertiary (0.156) and primary sector (0.052), which means the difference of the regional competitive effect in the secondary sector was biggest. Finally, the mean value of the regional competitive effect of the three sectors is 0.010, 0.043 and 0.018. This indicates that the regional competitive effect of the three sectors in RBCs all have positive effects on economic

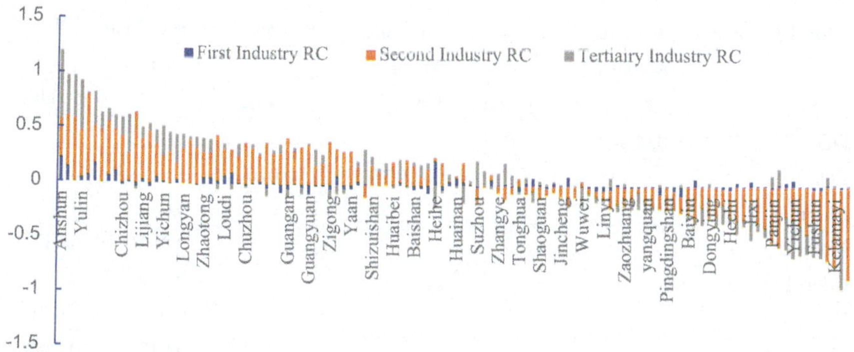

Fig. 4.5 Regional competitive effect of three industries

resilience, and the contribution of the secondary sector is bigger, which means that, just as expected, mining and resource processing industries are more competitive in RBCs.

In short, economic resilience is generally determined by the regional competitive effect in different regions, and industrial structure generally had negative effects on regional economic resilience in RBCs. Tertiary industry has positive effects, while the primary and secondary sectors had negative effect on industrial structure effect. Regional competitive of three sectors in RBCs all had positive effect on economic resilience, and the second sector had the highest contribution.

4.5.3 *Effect of Industrial Structure on Economic Resilience*

Since the economic resilience of RBCs show different features before and after 2012, we apply the panel regression model to examine the effect of industrial structure on resilience of three periods: 2008–2016, 2008–2012 and 2013–2016, shown in Tables 4.1, 4.2 and 4.3. In each period, we apply three models to explore the impact of different industrial structure indicators on resilience. Model 1 and 2 explore the effect of industrial diversification and specialization on resilience, and model 3 explore the effect of seven certain industries' specialization. To ensure the accuracy of the regression results, the testing of the model is performed in each model. First, we test the multicollinearity of indicators by variance inflation factor (VIF). Second, we use the F-test to decide whether to select pooled model or fixed effects model. Finally, we use the Hausman test to decide whether to select fixed effects model or random effects model.

From the regression results of industrial structure on economic resilience at different stages, we could find that the factors affecting economic resilience varied across different stages of the crisis. To the whole period, we may draw the following

Table 4.1 The Regression results of industrial structure on economic resilience, 2008—2016

	Model 1		Model 2		Model 3	
	Coef.	p-value	Coef.	p-value	Coef.	p-value
DIV	0.084***	0.000				
LQ_r			-0.00034**	0.010		
Mining LQ					−0.084***	0.000
Manu LQ					−0.252**	0.045
Const LQ					−0.050	0.224
Ret LQ					−0.049*	0.099
Finn LQ					−0.019	0.792
Tech LQ					−0.028***	0.000
Edu LQ					−0.202**	0.016
PCGDP	−1.91e–06*	0.096	−1.45e–06**	0.036	−1.51e–06	0.180
Pop_den	0.000494	0.468	−0.000192	0.160	0.0004397	0.558
FDI	0.147	0.140	0.173*	0.069	0.193**	0.027
Fix_Invest	0.055	0.462	−0.015	0.818	0.116*	0.072
Unemployment	−0.007*	0.072	−0.006*	0.055	−0.005	0.141
Constant	−0.119	0.611	0.328***	0.000	0.829***	0.008
Number of obs	1025		1025		1025	
R-squared	0.066		0.005		0.138	
F-test	F(113, 905) = 21.02, Prob > F = 0.0000		F(113, 906) = 20.75, Prob > F = 0.0000		F(113, 898) = 22.84, Prob > F = 0.0000	
Chi-square test	16.481**		7.564		31.525***	
Estimation model	Fixed effects model		Random effect model		Fixed effects model	

***p < 0.01, **p < 0.05, *p < 0.1

results. First, both diversification and specialization of industrial structure has significant effect on regional economic resilience at 5% level; DIV has positive effect, while LQr has negative effect. This indicates that RBCs with a more diversified structure have a higher ability to cope with the crisis, while specialized regions suffer more from recession. Second, the location quotient of seven selected industries all had negative effects on regional economic resilience; the mining, technical service and geologic prospecting had a significant effect at 1%, manufacturing, education at 5%, wholesale and retail at 10% level. We can find that RBCs with a high share of mining and related sector (technical service and geologic prospecting) have the greatest possibility of low resilience. Finally, from control variables, we can find that the regional development level and unemployment generally have a significant negative impact, which indicates that RBCs with a higher development level suffer more from the recession.

Table 4.2 The regression results of industrial structure on economic resilience, 2008—2012

	Model 1		Model 2		Model 3	
	Coef.	p-value	Coef.	p-value	Coef.	p-value
DIV	0.359**	0.032				
LQ_r			0.0001092	0.543		
Mining LQ					−0.010	0.396
Manu LQ					−0.192*	0.080
Const LQ					0.008	0.825
Ret LQ					−0.014	0.822
Finn LQ					0.228***	0.001
Tech LQ					−0.064	0.307
Edu LQ					−0.117*	0.081
PCGDP	2.39e-06***	0.002	2.66e–06***	0.004	2.95e–06***	0.001
Pop_den	−0.002	0.139	−0.0002353	0.105	−0.000154	0.305
FDI	−0.023	0.927	0.004	0.965	0.030	0.713
Fix_Invest	0.125**	0.049	0.127**	0.034	0.103	0.121
Unemployment	0.001	0.531	0.003	0.380	0.002	0.670
Constant	−0.201*	0.074	0.100	0.305	0.249	0.127
Number of obs	570		570		570	
R-squared	0.084		0.067		0.115	
F-test	F(113, 450) = 1.23, Prob > F = 0.0000		F(113, 450) = 21.60, Prob > F = 0.0000		F(113, 442) = 19.62, Prob > F = 0.0000	
Chi-square test	5.576		6.998		16.236	
Estimation model	Random effect model		Random effect model		Random effect model	

***p < 0.01, **p < 0.05, *p < 0.1

Table 4.2 reports the results of the regression analysis from 2008 to 2012, which show following results. First, only DIV, Manu LQ, Finn LQ and Edu LQ have significant effects on regional resilience. Second, the diversification of the industrial structure still has positive effects and its coefficient reaches the highest, while the effect of specialization was not significant. Third, the location quotient of manufacturing and education industries has a negative effect, while financial intermediation in this stage has a significant positive effect. This indicates interesting evidence that RBCs with high proportion of financial intermediation tend to have a high ability to cope with the recession. This may be attributed to the fact that the financial intermediation in RBCs is backward development, and that the financial intermediation firms in RBCs are mainly state-owned banks with an internally oriented economy, which is less influenced by the global crisis.

Table 4.3 reports the regression results between 2013 and 2016, from which we can identify several key features of the relationship between industrial structure and

Table 4.3 The regression results of industrial structure on economic resilience, 2013—2016

	Model 1		Model 2		Model 3	
	Coef.	p-value	Coef.	p-value	Coef.	p-value
DIV	0.067***	0.000				
LQ_r			−0.0001154	0.314		
Mining LQ					−0.031***	0.007
Manu LQ					−0.339***	0.000
Const LQ					0.106*	0.077
Ret LQ					0.006	0.825
Finn LQ					−0.069	0.304
Tech LQ					−0.010**	0.035
Edu LQ					0.132**	0.034
PCGDP	7.79e–06***	0.000	6.82e–06***	0.001	4.67e–06	0.000
Pop_den	0.0001215	0.839	0.000209	0.650	0.0000529	0.716
FDI	0.207**	0.033	0.340***	0.001	0.305	0.000
Fix_Invest	−0.018	0.859	−0.133	0.213	−0.023	0.796
Unemployment	−0.005**	0.019	−0.008***	0.003	−0.007	0.001
Constant	−0.418*	0.061	−0.181	0.314	0.015	0.912
Number of obs	455		455		456	
R-squared	0.322		0.199		0.110	
F-test	F(113, 335) = 32.81, Prob > F = 0.0000		F(113, 336) = 27.70, Prob > F = 0.0000		F(113, 329) = 25.09, Prob > F = 0.0000	
Chi-square test	27.674***		20.144***		21.33*	
Estimation model	Fixed effects model		Fixed effects model		Random effect model	

***$p < 0.01$, **$p < 0.05$, *$p < 0.1$

resilience. First, diversification has significant positive effect on economic resilience at 1% level; the specialization has a negative effect, but not significantly. Second, the resilience probability is increased by the construction and education specialization. After 2008, China introduced a more accommodative monetary policy, which directly promoted the rise in real estate prices. The real estate industry has become an important driver of economic growth, and it even became the leading industry in some RBCs. Furthermore, The Four Trillion Investment Plan mainly focused on infrastructural construction. All those activities have a certain time lag and construction has a positive effect. Finally, mining and its related sectors, technical service and geologic prospecting and manufacturing seem to decrease the regional resilience probability of RBCs. This may be highly related to China's de-capacity policy and the low international mineral resource prices.

In general, diversified industrial structures have significant positive effects on economic resilience, while specialization tends to lower the resilience probability of RBCs. The location quotient of mining, and its related industries, technical service and geologic prospecting and manufacturing, always has a negative effect on economic resilience. Therefore, fostering the diversification of industrial structure and reducing the proportion of the mining sector have become urgent tasks to improve economic resilience in RBCs in China. Moreover, the effect of other sectors varies at different stages of the crisis. At the beginning of the crisis, financial intermediation seems to increase the resilience probability of RBCs, and then the construction and education sectors begin to promote the economic resilience after 2012.

4.6 Results and Discussions

(1) *Resilience characteristics of RBCs with a specialized economy*

Compared with previous research which focused on a certain region, country or even continent, e.g., Europe (Breathnach et al., 2015; Brown & Greenbaum, 2017; Giannakis & Bruggeman, 2017a), our research explored the resilience of a sample of Chinese cities, i.e., the RBCs, and the effects of industrial structure on economic resilience. Although this research may not be representative for the whole territory of China, it can provide development reference for regions with a specialized economy.

Our results suggest that RBCs in China with a structurally weak economy had high economic resilience, especially at the beginning of the global financial crisis and the recession in RBCs was delayed. This finding does not support our first hypothesis. In contrast to what the existing literature argues, the specialized industrial structure of Chinese RBCs does not play a constraining role in shaping resilience. Moreover, the results also show great differences with Martin et al.'s (2016) UK-based evidence that industrial structure shapes the early processes of resilience. The novelty is that the industrial structure of Chinese RBCs seems to have less impact on the resistance and recoverability, but more on the post-recovery phase related to renewal and transformation.

Despite disadvantages of local structural preconditions, RBCs show higher resilience than expected. Three enabling factors can be identified: namely, sectorial, resourceful, and institutional. First, RBCs are single industry based and state-owned economies, which are highly embedded in national institutional-economic structures. Their main sectors are peripheral, less marketized and even very much self-contained, disconnected to the global economy. Therefore, compared to the export-led manufacturing dominated economy in many other cities of China, RBCs with their particular sectoral characteristics are able to avoid getting the direct hit from the economic crisis. Second, although the resource curse exists, we cannot deny the role of abundant mineral resources for the RBCs to withstand the economic crisis. Third, and perhaps the most important one, is that, compared to other cities, RBCs stand for higher political position in the national energy security system, they thus gain stronger legitimacy to draw more political attention from the central and provincial governments and receive a greater number of fiscal transfer and policy-led investment projects (e.g., industries, infrastructure and livelihood) when facing crises (Hu, 2017; Hu & Hassink, 2017a). More importantly, this institutional advantage also gives local and extra-local agents more power to initiate and support resilience building activities (Hu & Yang, 2019).

(2) *How industrial structure affects economic resilience in RBCs*

Within Chinese RBCs, our results suggest that most of RBCs' industrial structures had negative effects on economic resilience, especially with high proportions of the primary and secondary sector. This is consistent with the existing research on many peripheral and organizationally thin regions where the presence of systemically weak local industrial structure does negatively limit the potential of resilience (Hu & Hassink, 2020; Isaksen & Trippl, 2017; Martin et al., 2016). In this sense, our evidence shows that the dynamism of economic resilience was more affected by the agency effect than the industrial structure effect. That is to say, the role of region-specific factors, such as state agency and labor dynamics, have appeared to be prominent and thus require considerably more attention (Hu & Yang, 2019; Martin et al., 2016).

Although the effect of industrial structure on economic resilience of RBCs was variable, industrial diversification was the only factor that always had significantly positive effects at different stages of the crisis. This result supports our second hypothesis, and it does not contradict with previous findings that RBCs with a typical specialized economy had high economic resilience. Compared with other cities in China RBCs had higher resilience, while within RBCs cities with diversified industrial structures showed higher resilience. This provides an important orientation for policy making in RBCs to improve resilience and verifies the importance of industrial diversification with Chinese evidence. As Xiao et al. (2018) stressed, diversification is an important way in which the industrial base of an economy is renewed and broadened. Chinese RBCs with a more specialized industrial structure tended to have lower long-term economic resilience. This is consistent with Brown and Greenbaum (2017), who found out that specialized Ohio counties in the U.S. were less resilient in terms of their ability to absorb shocks.

In the end, our results suggest that RBCs specialized in manufacturing always had a significant negative impact on regional economic resilience. This echoes with previous studies on European cases where specialized manufacturing regions suffer more from workforce loss and global supply chain turbulence (Giannakis & Bruggeman, 2017a). Our analysis also suggests that the specialization of mining and its related sectors are negative for regional economic resilience. In the Chinese context, this may be related to the fact that employees working in state-owned enterprises with "*bianzhi*" (a permanent position in public and state owned institutions) cannot become redundant, and local governments/state-owned enterprises have to bear social burdens which undermines resilience in Chinese RBCs (Tan et al., 2017). Moreover, this can be also explained by the long-term lock-in effects derived from over-specialization in mining sectors which inevitably weaken resilience (Hu & Hassink, 2017b).

(3) *Policy implications*

The understanding of the nature of regional economic resilience and of how the industrial structure affects it can help to generate four policy implications. First, more attention should be paid to the place-specific agency effects, since it provides dynamism and opportunities for economic resilience (Sedita et al., 2017). Region-specific comparative factors, such as institutional strengths and policy capabilities related to state agency, deserve more policy attention (Bristow & Healy, 2014). Second, RBCs need to foster and diversify the tertiary sector, as it can not only increase resilience by creating more jobs but also sustain resilience by absorbing labor. Third, RBCs may not be able to achieve diversification in the short term, due to the lock-in effects in resource specialization. Therefore, the focus on exploiting comparative advantage, particularly by the bricolage and mobilization of non-structural assets across different scales for unrelated new regional industrial path development, is the key to strengthen resilience (Isaksen & Trippl, 2017; Trippl et al., 2020). Finally, since resilience implementation involves cooperation and collaboration between government agencies, non-governmental organizations and private businesses, detailed policy-making should combine top-down, bottom-up, and mixed approaches, and properly handle the relationship between regional industrial structure and multi-scalar agency (Chmutina et al., 2016; Shamsuddin, 2020; Zhang et al., 2019).

4.7 Conclusions

This paper quantitatively explores economic resilience after the global economic crisis of RBCs in China. It analyses how industrial structure affects resilience by decomposing the resilience and by a panel regression. It contributes to the role of industry structure in affecting resilience and draws four key conclusions. First, Chinese RBCs with a structurally weak economy had high resilience, especially at the beginning of the global financial crisis. They encountered a deep recession after 2012, which shows that the recession in RBCs was delayed. Second, industrial structure had negative effects on regional economic resilience in RBCs, indicating that the agency effect is the key determinant of resilience among different RBCs. Third, diversification clearly increase the resilience, but specialization tends to decrease the resilience in RBCs. The specialization in mining, in particular, has negative effects on resilience. Finally, the effect of specialization of certain industries varies at different stages of the crisis. Although some works have been done in this paper, but still, there are several inadequacies and possible extensions for future research. This paper quantitatively identified the role of industrial structure in affecting resilience, but the reasons behind it deserve further clarification through more in-depth qualitative research on the role of dynamic agency and institutions, as well as their interaction with industrial structure (Bristow & Healy, 2014; Evenhuis, 2017; Hassink & Gong, 2020; Hu & Hassink, 2020). Besides, analyses on other indicators of economic and social structure, including relatedness, knowledge bases, networks, institutional thickness, culture and norms, social capital are required for future research. To end up, this paper suggests that regional economic resilience research should avoid the narrow "structure matters" view by embracing a multi-scalar and multi-actor, and contextually sensitive perspective incorporating the role of structure and agency in a co-evolutionary manner (Gong & Hassink, 2019; Martin & Sunley, 2020). We argue the interplay between structure (assets, power and elements) and agency (actions, actors and ability) in a broader context is the key to understand the determinants of regional economic resilience over time and space (Hu & Hassink, 2020). This should be put at the core in future regional economic resilience research in China and beyond.

Acknowledgements Juntao Tan and Robert Hassink have made great contributions to this chapter. We express our heartfelt thanks to them.

References

Balland, P. A., Rigby, D., & Boschma, R. (2015). The technological resilience of US cities. *Cambridge Journal of Regions, Economy and Society, 8*(2), 167–184.

Boschma, R. (2015). Towards an evolutionary perspective on regional resilience. *Regional Studies, 49*(5), 733–751.

Breathnach, P., Van Egeraat, C., & Curran, D. (2015). Regional economic resilience in Ireland: The roles of industrial structure and foreign inward investment. *Regional Studies, Regional Science, 2*(1), 497–517.

Bristow, G., & Healy, A. (2014). Regional resilience: An agency perspective. *Regional Studies, 48*(5), 923–935.

Bristow, G., & Healy, A. (2018). Innovation and regional economic resilience: An exploratory analysis. *The Annals of Regional Science, 60*(2), 265–284.

Bristow, G., & Healy, A. (2020). *Introduction to the handbook on regional economic resilience.* Edward Elgar Publishing.

Brown, L., & Greenbaum, R. T. (2017). The role of industrial diversity in economic resilience: An empirical examination across 35 years. *Urban Studies, 54*(6), 1347–1366.

Cainelli, G., Ganau, R., & Modica, M. (2019). Industrial relatedness and regional resilience in the European Union. *Papers in Regional Science, 98*(2), 755–778.

Di Caro, P. (2015). Recessions, recoveries and regional resilience: Evidence on Italy. *Cambridge Journal of Regions, Economy and Society, 8*(2), 273–291.

Chmutina, K., Lizarralde, G., Dainty, A., & Bosher, L. (2016). Unpacking resilience policy discourse. *Cities, 58*, 70–79.

Christopherson, S., Michie, J., & Tyler, P. (2010). Regional resilience: Theoretical and empirical perspectives. *Cambridge Journal of Regions, Economy and Society, 3*, 3–10.

Courvisanos, J., Jain, A., & Mardaneh, K. (2016). Economic resilience of regions under crises: A study of the Australian economy. *Regional Studies, 50*(4), 629–643.

Dawley, S. (2014). Creating new paths? Offshore wind, policy activism, and peripheral region development. *Economic Geography, 90*(1), 91–112.

Doran, J., & Fingleton, B. (2016). Employment resilience in Europe and the 2008 economic crisis: Insights from micro-level data. *Regional Studies, 50*(4), 1–13.

Economist (2014). The north-east: Back in the cold. The Economist.

Eraydin, A. (2016). Attributes and characteristics of regional resilience: Defining and measuring the resilience of Turkish regions. *Regional Studies, 50*(4), 600–614.

Evenhuis, E. (2016). *The political economy of adaptation and resilience in old industrial regions: A comparative study of South Saarland and Teesside*. Newcastle University.

Evenhuis, E. (2017). New directions in researching regional economic resilience and adaptation. *Geography Compass, 11*, 1–15.

Ezcurra, R., & Rios, V. (2019). Quality of government and regional resilience in the European Union. Evidence from the Great Recession. *Papers in Regional Science, 98*(3), 1267–1290.

Faggian, A., Gemmiti, R., Jaquet, T., & Santini, I. (2018). Regional economic resilience: The experience of the Italian local labor systems. *The Annals of Regional Science, 60*(2), 393–410.

Fingleton, B., Garretsen, H., & Martin, R. (2012). Recessionary shocks and regional employment: Evidence on the resilience of UK regions. *Journal of Regional Science, 52*(1), 109–133.

Fornahl, D., Hassink, R., Klaerding, C., Mossig, I., & Schröder, H. (2012). From the old path of shipbuilding onto the new path of offshore wind energy? The case of Northern Germany. *European Planning Studies, 20*(5), 835–855.

Fredin, S., Miörner, J., & Jogmark, M. (2019). Developing and sustaining new regional industrial paths: Investigating the role of "outsiders" and factors shaping long-term trajectories. *Industry and Innovation, 26*(7), 795–819.

Fröhlich, K., & Hassink, R. (2018). Regional resilience: A stretched concept? *European Planning Studies, 26*(9), 1763–1778.

Giannakis, E., & Bruggeman, A. (2017a). Determinants of regional resilience to economic crisis: A European perspective. *European Planning Studies, 25*(8), 1394–1415.

Giannakis, E., & Bruggeman, A. (2017b). Economic crisis and regional resilience: Evidence from Greece. *Papers in Regional Science, 96*(3), 451–476.

Gong, H., & Hassink, R. (2019). Co-evolution in contemporary economic geography: Towards a theoretical framework. *Regional Studies, 53*(9), 1344–1355.

Grillitsch, M., & Sotarauta, M. (2019). Trinity of change agency, regional development paths and opportunity spaces. *Progress in Human Geography*, 1–20.

Guan, H., Lin, W., Zhang, P., Lo, K., Li, J., & Li, L. (2018). Analyzing industrial structure evolution of old industrial cities using evolutionary resilience theory: A case study in Shenyang of China. *Chinese Geographical Science, 28*(3), 516–528.

Hassink, R. (2010). Regional resilience: A promising concept to explain differences in regional economic adaptability? *Cambridge Journal of Regions, Economy and Society, 3*(1), 45–58.

Hassink, R., & Gong, H. (2020). Regional resilience. In A. Kobayashi, M. Boyle, N. Castree, F. Collins, J. Crampton, S. de Leeuw, J. Tyner (Eds.), *International encyclopedia of human geography* (pp. 351–355). Elsevier.

Hervas-Oliver, J. L., Jackson, I., & Tomlinson, P. R. (2011). "May the ovens never grow cold": Regional resilience and industrial policy in the north Staffordshire ceramics industrial district–with lessons from Sassoulo and Castellon. *Policy Studies, 32*(4), 377–395.

Holling, C. S. (1973). Resilience and stability of ecological systems. *Annual Review of Ecology and Systematics, 4*, 1–23.

Hu, X. (2017). From coal mining to coal chemicals? Unpacking new path creation in an old industrial region of transitional China. *Growth and Change, 48*(2), 233–245.

Hu, X., & Hassink, R. (2017a). Exploring adaptation and adaptability in uneven economic resilience: A tale of two Chinese mining regions. *Cambridge Journal of Regions, Economy and Society, 10*(3), 527–541.

Hu, X., & Hassink, R. (2017b). New perspective on restructuring of old industrial areas in China: A critical review and research agenda. *Chinese Geographical Science, 27*(1), 110–122.

Hu, X., & Hassink, R. (2020). Adaptation, adaptability and regional economic resilience: A conceptual framework. In G. Bristow & A. Healy (Eds.), *Handbook on regional economic resilience* (pp. 54–68). Edward Elgar.

Hu, X., & Yang, C. (2019). Institutional change and divergent economic resilience: Path development of two resource-depleted cities in China. *Urban Studies, 56*(16), 3466–3485.

Isaksen, A., & Trippl, M. (2017). Exogenously led and policy-supported new path development in peripheral regions: Analytical and synthetic routes. *Economic Geography, 93*(5), 436–457.
Kemeny, T., & Storper, M. (2015). Is specialization good for regional economic development? *Regional Studies, 49*(6), 1003–1018.
Kitsos, A., Carrascal-Incera, A., & Ortega-Argil.s, R. (2019). The role of embeddedness on regional economic resilience: Evidence from the UK. *Sustainability, 11*(14), 3800.
Li, H., Lo, K., & Wang, M. (2015). Economic transformation of mining cities in transition economies: Lessons from Daqing, Northeast China. *International Development Planning Review, 37*(3), 311–328.
Martin, R. (2012). Regional economic resilience, hysteresis and recessionary shocks. *Journal of Economic Geography, 12*(1), 1–32.
Martin, R., & Gardiner, B. (2019). The resilience of cities to economic shocks: A tale of four recessions (and the challenge of Brexit). *Papers in Regional Science, 98*(4), 1801–1832.
Martin, H., & Martin, R. (2017). Policy capacities for new regional industrial path development: The case of new media and biogas in southern Sweden. *Environment and Planning C, 35*(3), 518–536.
Martin, R., & Sunley, P. (2015). On the notion of regional economic resilience: Conceptualization and explanation. *Journal of Economic Geography, 15*(1), 1–42.
Martin, R., & Sunley, P. (2020). *Regional economic resilience: Evolution and evaluation.* Edward Elgar Publishing.
Martin, R., Sunley, P., Gardiner, B., & Tyle, P. (2016). How regions react to recessions: Resilience and the role of economic structure. *Regional Studies, 50*(4), 561–585.
Martin, R. L. (2018). Shocking aspects of regional development: Towards an economic geography of resilience. In G. Clark, M. Gertler, M. P. Feldman, & D. W. Jcik (Eds.), *The new Oxford handbook of economic geography* (pp. 839–864). Oxford University Press.
Safford, S. (2009). *Why the garden club couldn't save Youngstown: The transformation of the Rust Belt.* Harvard University Press.
Sedita, S. R., De Noni, I., & Pilotti, L. (2017). Out of the crisis: An empirical investigation of place-specific determinants of economic resilience. *European Planning Studies, 25*(2), 155–180.
Sensier, M., Bristow, G., & Healy, A. (2016). Measuring regional economic resilience across Europe: Operationalizing a complex concept. *Spatial Economic Analysis, 11*(2), 1–24.
Shamsuddin, S. (2020). Resilience resistance: The challenges and implications of urban resilience implementation. *Cities, 103*, 102763.
Simmie, J., & Martin, R. (2010). The economic resilience of regions: Towards an evolutionary approach. *Cambridge Journal of Regions, Economy and Society, 3*(1), 27–43.
Tan, J., Lo, K., Qiu, F., Liu, W., Li, J., & Zhang, P. (2017). Regional economic resilience: Resistance and recoverability of resource-based cities during economic crises in Northeast China. *Sustainability, 9*, 2136.
Trippl, M., Saumgartinger-Seiringer, S., Frangenheim, A., Isaksen, A., & Rypestl, J. O. (2020). Unravelling green regional industrial path development: Regional preconditions, asset modification and agency. *Geoforum, 111*, 189–197.
Webber, D. J., Healy, A., & Bristow, G. (2018). Regional growth paths and resilience: A European analysis. *Economic Geography, 94*(4), 355–375.
Xiao, J., Boschma, R., & Andersson, M. (2018). Industrial diversification in Europe: The differentiated role of relatedness. *Economic Geography, 94*(5), 514–549.
Zhang, Y., Long, H., Ma, L., Tu, S., Liao, L., Chen, K., & Xu, Z. (2019). How does the community resilience of urban village response to the government-led redevelopment? A case study of Tangjialing village in Beijing. *Cities, 95*, 102396.
Zhang, W., Yu, J., & Wang, D. (2011). Study on the development path and pattern of substituted industry in resource-based cities. *Bulletin of the Chinese Academy of Sciences, 26*, 134–141. (in Chinese).
Zhu, S., He, C., & Zhou, Y. (2017). How to jump further and catch up? Path-breaking in an uneven industry space. *Journal of Economic Geography, 17*(3), 521–545.

Chapter 5
What Matters for Regional Economic Resilience Amid COVID-19? Evidence from Cities in Northeast China

5.1 Introduction

Over the past two decades, cities have been increasingly exposed to various risks and crises from multiple geographical scales (Giannakis & Bruggeman, 2017a). The 2008 global financial crisis, in particular, has generated a great scholarly concern on how and to what extent regional economies are able to recover from the crisis (Foster, 2007; Hill et al., 2008). Since then, regional resilience has been a buzzword in not only in academia but also in public consciousness, and intensively being equipped as a policy aim or agenda for regional development. The notion of resilience has been widely used in physics, disaster science, ecology and psychology, holding with its equilibrium idea of "bouncing back" or recovery (Hu & Hassink, 2017a). However, in economic geography, resilience is rather considered as a multi-actor involved, contextualized and non-equilibrium process. It involves with both reactive and proactive agency and institutions that may not only withstand or absorb shocks but also seek for "bouncing forward" for further adaptability and growth (Bristow & Healy, 2014; Martin, 2012).

Recently, there is a growing consensus that regional economic resilience has a malleable and never-ending characteristic and thus is an inherently shock-prone process of development (Martin, 2018). Resilience essentially represents the very core feature of regional economic evolution in an uncertain context: a never-ending process in which actors and institutions prepare, anticipate and adapt to frequent nature and man-made challenges for regional sustainable growth. For economic geographers, the key research theme is to explain why regional economies differ in resilience to varied shocks and what determines such differential impacts over time and space (Hassink, 2010; Hu & Hassink, 2020; Martin, 2012; Martin et al.,

(This chapter is a reprint version from: Hu, X., Li, L., & Dong, K. (2022). What matters for regional economic resilience amid COVID-19? Evidence from cities in Northeast China. Cities, 120, 103440.)

X. Hu, *Uneven Economic Resilience of Old Industrial Cities in China*,
Urban Sustainability, https://doi.org/10.1007/978-981-99-9279-9_5

2016). On the topic, three main inter-connected research trends can be identified. First, resilience has been increasingly conceived as a multi-dimensional process that is not subject to shock-centered recovery and resistance. Its current epistemological development emphasizes on the pre-shock vulnerability or sensitivity of a regional economy with social and institutional characteristics involved, and on its adaptability for new industrial path development during and after shocks (Martin & Sunley, 2015). Second, both quantitative and qualitative studies have suggested that resilience is not about something you have but more about something or the ways in which you react. In other words, although a region's historically inherited structural elements and assets can influence the scope and ability of agency to shocks, the role of human agency is regarded as one of most crucial factors in making sense of change and resilience (Bristow & Healy, 2014, 2020; Hu & Yang, 2019). Third, the nature of shocks is itself of critical importance as a point of departure to understand regional resilience. Arguably, the specificities of shock, such as its socio-spatial characteristics of origin, impact of scale, duration, depth and target population, can fundamentally define resilience of what, to whom, and for where and what. Moreover, shocks, even if a one-off event, may generate long-lasting or permanent effects (Martin, 2018). Regional resilience thus is about the dynamism and adaptability of regions to constantly on-going and changing pressure (Martin et al., 2016). It indeed embodies collective properties and dynamic capabilities of regions underpinned by the multi-scalar interplay between structure, agency and context (David, 2018; Hu & Hassink, 2020; Hu & Yang, 2019; Tan et al., 2020a).

The 2019 coronavirus disease (COVID-19) pandemic, as an on-going global pandemic continually spreading across the world, has led to a truly worldwide crisis. As of the time of writing in May 2021, over 160 million cases of COVID-19 from 188 countries and territories have been reported, causing more than 3.32 million deaths (JHU, 2021). Since COVID-19 is spread between people in close physical proximity, human activities of any kind as long as requiring mass gatherings are strictly restricted. Only for months, the pandemic has led to the largest global economic disruption since the Great Depression (Nicola et al., 2020). Given such an unprecedented scale and impact depth of the crisis, the resilience idea has received a renewed deal of research attention.

Many scholars have dealt with exploring the resilience mechanisms and measures in psychological recovery of particular groups and workforces, risk governance of global supply chains, and system adjustment of certain sectors (food, healthcare, etc.) (Chen & Bonanno, 2020; Ivanov & Dolgui, 2020; Klassen & Murphy, 2020). A growing number of researchers have started to understand the geographically uneven impact and consequences of the pandemic. Different nations, in particular, are evidently found to have a wide variation with regard to the spatiality of transmission, vulnerabilities of human life, qualities of healthcare systems, effectiveness of state policy/measure to contain it, and subsequent socio-economic consequences (Åslund, 2020; Asongu et al., 2020; Chung et al., 2020; Rose-Redwood et al., 2020; Wang et al., 2020). Others argue that COVID-19 itself mirrors and produces more inequalities of power and space, as well as socio-political conflicts within and among nations, regions, localities (Dodds et al., 2020; also see the Special

Issues in Dialogues in Human Geography, volume 20 Issue 2, and in Eurasian Geography and Economics, volume 61, Issue 4–5.). It is suggested that the state at multiple spatial levels has emerged as a key and powerful territorial agency on an unprecedented scale—through controlling global travel restrictions, domestic lock-down, social distancing and other measures—in creating an unconventional movement of geopolitics and socio-spatial governance (Chan et al., 2020; Grundy-Warr & Lin, 2020). These studies have emphasized more on how the pandemic has been generating geographies of governance quality and performance mainly at the national level (Asongu et al., 2020). However, and surprisingly, the topic of how regional economies response and adapt to the crisis—related to regional economic resilience—has not yet become to the fore in the COVID-19 literature (OECD, 2020). Several questions are raised that deserve more attention in both academia and policy spheres: How regional and city economies resist and respond to COVID-19? What impact factors shape the geographies of regional economic resilience under COVID-19? What kind of useful policy implications, if any, can enhance regional resilience against COVID-19 and crises of that kind in near future?

This paper thus aims to investigate the determinants of regional economic resilience under COVID-19, particularly by providing evidence on how city-level economies of Northeast China respond to the crisis at the early stage. This study is organized with five sections as follows. Based on the literature review on factors shaping regional resilience under the 2008 financial crisis, Sect. 5.2 provides hypotheses about what factors might matter for regional resilience to COVID-19. Section 5.3 introduces study area, methodology and data. After showing results in Sects. 5.4 and 5.5 discusses the determinants of uneven regional resilience in Northeast China under COVID-19. Section 5.6 concludes.

5.2 Literature Review and Hypotheses

5.2.1 What Affect Regional Economic Resilience Under the 2008 Financial Crisis

The 2008 financial crisis heralded the most severe economic downturn in the history of major developed countries, and also brought about the slowdown of growth rate in many emerging economies such as China (Bristow & Healy, 2020; Hu & Hassink, 2017a). In the past decade, one of the most intriguing questions in economic geography is why some regions are more able to cope with the crisis than others, and what shapes their abilities to recover and also adapt to uncertainties (Balland et al., 2015). In line with the theorizing of evolutionary economic geography (EEG), the notion of resilience has emerged as a key idea for explaining geographically uneven resilience of places to the crisis (Hassink, 2010). Three EEG conceptual lens for understanding what influences resilience can be distinguished.

First, the 'panarchy' idea and resilience. Inspired by ecological science, the 'adaptive cycle' model is initially adopted to understand regional resilience (Simmie & Martin, 2010). Resilience as process follows a sequential cycle of exploitation (emergence), conservation (stabilization), release (decline) and reorganization (restructuring). In this conceptualization, accumulated assets and firm connectedness shape the potential of a region's resilience. Given this firm-centric endogenous thinking, the model assumes market-led innovation, learning, and technological change are the key drivers for regional resilience (Martin, 2012). This however is not unproblematic. Multi-scalar state and public policies are also of utmost importance for resilience, which is largely neglected in the model (Hassink, 2010; Kakderi & Tasopoulou, 2017).

Second, path dependence/creation and resilience. A path dependence approach to resilience is precisely concerned with resistance and recovery (Simmie & Martin, 2010). It focuses on the adaptation within the "elasticity threshold" of a region's economy (Martin, 2012). Adaptation refers to a particular form of resilience trapped in the history, conditioned by a region's "portfolio", such as economic structure, regime setting, institutional forms and entrepreneurial culture (Martin & Sunley, 2015). However, a path creation approach defines resilience as the adaptability of pushing a system beyond its elasticity threshold and creating a new development path, via reorientation and renewal (Martin, 2012). Influenced by the canonical path dependence model, scholars argue that there is a trade-off between adaptation and adaptability (Grabher & Stark, 1997; Pike et al., 2010). Regions with related variety, loosely coupled firm networks and enabling institutional settings are resilient as they can overcome the trade-off (Boschma, 2015). Others, however, stress that adaptability is more influenced by state agency in broader geographical political economy (Pike et al., 2010). This is particularly the case for less favored regions: crisis can confer more political legitimacy for top-down policy interventions for adaptability (Hu & Hassink, 2017a). Their resilience therefore shows strong reliance on agency over structure (Hu, 2015; Hu & Yang, 2019; Tan et al., 2020a).

Third, complexity adaptive system theory (CAS) and resilience. More recently, the systems ontology has been emphasized in the analysis of regional economic resilience (Welsh, 2014). Bristow and Healy (2014) argue that real world economies are highly complex, non-linear and adaptive human systems. In other words, resilience is about adaptive capabilities collectively constructed by a wider range of agents confronted with certain crisis or shocks. The interpretations of risks and contexts, visions and agenda-setting for the future, ways of responses and policy decisions—in collective forms of agency and governance—can significantly shape the potential, geographies and trajectories of resilience (Bristow & Healy, 2014, 2020; Chung et al., 2020; Hu & Hassink, 2020).

Moreover, studies have further pointed out that inherent and inherited structural factors (e.g., industrial structure, openness, institutional settings) primarily matter for vulnerability and resistance, while 'adaptable' agency-based factors (e.g., business confidence, policies, leadership) precisely shape recoverability and reorientation (Martin et al., 2016; Martin & Sunley, 2020; Tan et al., 2020a). In this sense, resilience is also about context- and place-sensitive agency that adaptively co-evolve

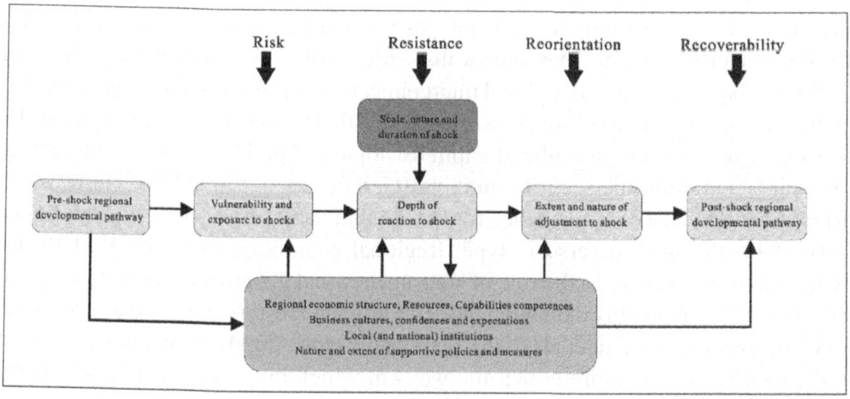

Fig. 5.1 Dimensions of regional resilience to economic shocks. *Source* Martin et al. (2016)

with existing historically conditioned structural elements for development. More recently, as Fig. 5.1 shown, the nature of risks is regarded as a key factor shaping resilience (Martin & Sunley, 2015). For example, the source/type of crisis (man-made or natural) matters for resilience to what. The impact object of a shock can determinate resilience of what and for whom (Martin, 2018). The duration of risk (one-off or short-term shock or slow-burn crisis) would condition the purpose of resilience (resilience for what) (Martin et al., 2016). Moreover, the impact of crisis may shift over space and time, becoming a key factor that creates uneven contextual conditions and contingencies for regional economic resilience (Bailey et al., 2020; Chung et al., 2020). In sum, resilience involves a crisis-induced process of co-evolution between structure and agency, constantly shaped by history, place and context.

5.2.2 *Hypotheses: Potential Factors Affecting Regional Resilience Under COVID-19*

The 2008 financial crisis and the COVID-19 crisis are bound to differ at least in terms of target, severity, scope and duration of impact they both have generated huge impact on the economy though. Three key differences with respect to the resilience theme between the two crises can be identified. Firstly, while the key concern of the financial crisis is about the financial system recovery and demand-side economic growth, the COVID-19 pandemic is a public-health crisis that affects almost all sort of human activities across the world (OECD, 2020). The first priority response linking to resilience is to resist to and contain the spread of COVID-19. In this regard, government-led containment measures—city lockdowns, border controls, business/factory close-down, transportation bans, and social distancing retractions—are indeed powerful forces that affect the real economy overwhelmingly, both on the production and demand side across nearly all sectors. Second, traditional regional structural advantages might lose advantages for economic resilience under COVID-19. For instance, evidence has shown that cities with dense market clustering and workforce base, or with wider global interconnections in supply chain have exhibited higher economic vulnerability (Lawreniuk, 2020). This has led to a critique on the existing redundant and neo-liberal capitalist logics of global supply chain lacking robustness and domestic security thinking (Gereffi, 2020; Liu, 2020). Third, while firms are the key actors of resilience for the 2008 financial crisis, but the actors under COVID-19 are much diverse in type. Regional economies under COVID-19 are extraordinarily sensitive to the role of state agency and governance in containing the pandemic (also inevitably damaging the economy) and restoring the social-economic order afterwards (Asongu et al., 2020; Garavaglia et al., 2020). A shared recognition in the COVID-19 literature is that the ways in which the governance (particularly the efficacy of state responses and measures) to the COVID-19 matters most the geographies of crisis impacts and consequence (He et al., 2020).

Based on the above review and discussions, we assume that both structure and agency affect regional economic resilience under COVID-19. More importantly, the latter, particularly state agency in controlling the trajectory of economic recovery and adaptation will be given more attention. Given further review on the COVID-19 literature, four key hypotheses on what might shape regional economic resilience in the crisis are proposed as follows.

(1) *Basic industrial composition*

Basic industrial composition here refers to the proportions of the key economic sectors in the gross product of a regional economy (Atikian, 2013). The COVID-19 pandemic entails a much wider socio-economic crisis than the economic one does. The priority of resilience thus is not economic, but rather it is more related to protect human lives at the cost of economic development. That means, any forms of human socio-economic activities with physical gatherings are restricted, in order to minimize the risk of exposure to COVID-19. This has inevitably led to disruptions and decline to almost all of the supply-side and demand-side sectors. In particular, the labor-intensive, export-led secondary industry amongst many countries and regions has been severely hit by the crisis (Nicola et al., 2020). In the first two months after the COVID-19 outbreak, industrial production by major enterprises in China, for instance, dropped by 15.7 percent year on year (NBSC, 2020). Meanwhile, the tertiary industry has experienced a sharp drop in output, as a result of the implementation of human mobility control, real business shutdown and public activity restriction against COVID-19. We thus hypothesize:

H1: the share of secondary and tertiary industry in the whole economy strongly matters for regional economic residence. Cities with a higher share of secondary and tertiary industry in GDP present weaker resistance to COVID-19.

(2) *Industrial structure*

Industrial structure plays a crucial role in shaping a region's economic sensitivity and resistance to shocks (Martin, 2012). It is quantitatively evident that an industrially diversified or varied economic structure often provides greater regional resistance than does a specialized structure. Because the former can act as "shock absorber" to effectively disperse the impact of shocks (Angulo et al., 2018; Ringwood et al., 2019). In terms of diversification, scholars further distinguish the role of related variety and unrelated variety in affecting regional economic resilience (Boschma, 2015). It is found that, at least in a short run, a structure of unrelated variety is more able to withstand shocks. This is due to the heterogenous impacts of shocks to different sectors, suggesting that the strategy of "not to put all eggs in one basket" can enhance a region's resistance (Zhu et al., 2017). A related variety structure may be beneficial for long term resilience as it can promote regional positive externalities and innovation, while such structure may receive greater hit on the whole economic system in a short term. Because it helps the impact of shock to quickly spread from one specific industry to adjacent ones, which is not conducive to regional resistance (Content & Frenken, 2016).

We suggest that the role of industrial structure as aforementioned is also of highly relevance for regional economic resilience (resistance) under COVID-19. Although COVID-19 has generated a huge shock to the whole economy, it is arguably will not be able to impact all sectors negatively and equally. For instance, COVID-19 is more likely to affect sectors replying on the physical proximity and clusters of labor forces. It would have a limited impact on the virtual economy empowered by, for instance, Internet technologies (Gereffi, 2020). Furthermore, cities with an unrelated variety structure, if involving a mix of different industries/technologies, may have more resistance to COVID-19. We thus hypothesize:

H2: Industrially diversified cities in general provide greater economic resistance than cities with a specialized structure.

H3: An unrelated variety structure enables regional resistance.

(3) *Economic openness*

In the existing literature, regions with a high degree of economic openness often refer to economically advanced and globalized economies. They are able to attract a large number of resources, technologies and firms, and mobilize global assets for regional development with high efficiency, which in turn can improve the ability to cope with crisis (Eraydin, 2016; Giannakis & Bruggeman, 2017b; Martin & Sunley, 2015). However, the validity of this assumption is heavily dependent on the nature and scope of a crisis. If it affects international logistics and transport, or cause restrictions on export and import in a national or global scale, regions with higher economic openness rather face more difficulties and stresses (Martin & Sunley, 2020). COVID-19 is a truly global crisis to globalization, as it has caused massive production stoppages, port closure and international shipping suspension in the world's major globalized economies (e.g., EU, Japan, China, and USA). Regional economies with more global connections are likely to show greater uncertainty and vulnerability to COVID-19 (Nicola et al., 2020). The pandemic has even raised a critical rethink of the value of the existing global production networks and suggested to build up a more risk-averse national supply chains, or to create a new global sourcing incorporating inward-looking regional economies that are less-affected by COVID-19 (Bryson & Vanchan, 2020; Gereffi, 2020). In this regard, we hypothesize:

H4: Cities with a higher degree of economic openness are more vulnerable to COVID-19, showing weaker economic resilience.

(4) *Government control measures*

Besides the aforementioned structural factors, the role of state agency in affecting regional resilience has been increasingly highlighted particularly when facing global crisis (Ezcurra & Rios, 2019). The nature of a crisis (e.g. source, duration, scope, and impact object) would play a crucial role in underpinning the ways in which state actions unfold to cope with it (Martin, 2018). Unlike the 2008 economic crisis that mainly involves firm agency, COVID-19 as a world-wide pandemic crisis, government bears the first accountability to contain the virus and protect people's lives. City governments in China, for instance, have played a leading role in taking a series of strict measures to contain the spread of the pandemic, including "restricting access and entry for residential communities", "delaying the commencement of work" and "suspending public transportation and schooling", and even "city-scale lock-down" (Chen et al., 2020). These measures have inevitably brought the economy to a standstill. However, such standstill is temporary, particularly with the virus being increasingly contained in the first quarter of 2020 in China. In this sense, the relaxing or cancellation of these government control measures would help to restore basic economic activity and recovery the economy. We thus hypothesize:

H5: Government measures to contain COVID-19 matter significantly for regional economic resistance. State agency involving the cancellation of these restrictions, in particular, promotes regional recovery to COVID-19.

5.3 Study Area, Methodology and Data

5.3.1 Why Cities in Northeast China?

In this paper, cities in Northeast China were selected as the study area (Fig. 5.2), with two main reasons. On the one hand, the region's featured economic structure deserves regional resilience research in new crisis contexts (Hu & Yang, 2019; Tan et al., 2020a). It is evident that the region's economy was less sensitive to the 2008 financial crisis (Tan et al., 2020b) due to its long-term domination of state-owned enterprises characterized by inward-looking national supply chains and markets. Despite higher resistance, the structurally weak (over-specialization on mature heavy manufacturing and resource mining) economy has led to a weak recoverability and dynamism (Li et al., 2019; Tan et al., 2017). In this regard, exploring how COVID-19 affects Northeast China can allow us to advance the complex nature of regional

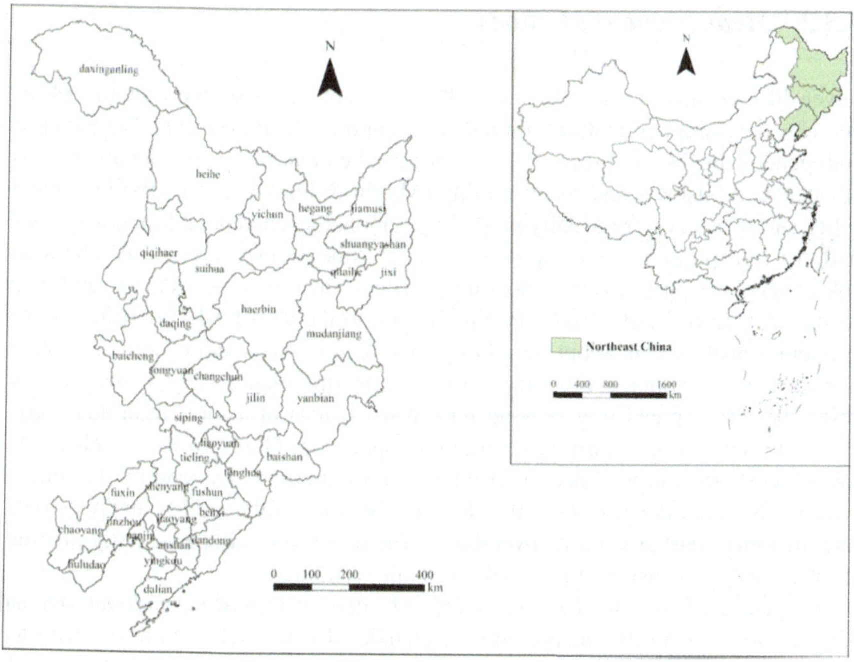

Fig. 5.2 Spatial distribution of cities in old industrial base in Northeast China

economic resilience. On the other hand, much research has emphasized the heterogeneity of resilience among Northeast cities, mainly shaped by regionally different industrial bases and configurations (Hu & Hassink, 2017a). However, the role of agency in shaping regional resilience is less examined in the existing literature. Arguably, human agents and their abilities to contain the pandemic spread on the one hand, and to minimize the negative impact on their economies on the other, are both essential to resilience.

Given this, in total 34 prefecture-level cities (expect Yanbian Korean Autonomous Prefecture and Daxinganling Area) were selected as the research unit for exploring the question of what matters for regional resilience under COVID-19. In the first quarter of 2020, the GDP growth rates of Heilongjiang, Jilin and Liaoning provinces were −8.3, −6.6 and −7.7%, respectively. 94% of the cities experienced negative economic growth. The national average GDP growth rate was −6.8%, and the economic development of Heilongjiang and Liaoning was more severely affected by the COVID-19. Compared with the first quarter of 2019, the GDP growth rates of Heilongjiang, Jilin and Liaoning fell by 257, 375, and 226%, respectively. The regional economy was affected by the COVID-19 and developed slowly.

5.3.2 Measurement Methods

Given different nature of shocks, research period of time, as well as availability of data sources, measurement methods on regional economic resilience vary. Nevertheless, four general attributes of regional resilience can be identified, namely, vulnerability, resistance, robustness and recoverability (Martin & Sunley, 2015). While vulnerability and resistance are usually shaped by regionally inherent and inherited assets and structural properties exposure to shocks, robustness and recoverability refer to the role of agency in purposively responding to shocks, in order to recover from and adapt to shocks (Martin et al., 2016). Putting them into the COVID-19 context, regional economic resilience is about the ability of a region—its socio-economic systematic capacities, resource mobilization power, institutional arrangements etc.,—to resist the virus spread (e.g., saving human lives first) in a short term and recovering the regional economy (e.g., business operation, GDP growth, employment, investment) afterwards. Since the pandemic has not come to an end by the time of writing, the economic resilience thus is primarily about vulnerability and resistance, although there partially has recoverability. The paper thus will particularly measure the resistance dimension of regional economic resilience.

To quantitatively reflect the resistance of cities, we construct a counterfactual function to compare the actual amount change of urban GDP with the expected change (Martin et al., 2016). The calculation formula of urban expected economic output change is as follows:

$$\left(\Delta R_i^{t+k}\right)^{expected} = \sum_j^n R_{ij}^t \cdot G_n^{t+k} \tag{5.1}$$

where R_{ij}^t is the output value of industry j in region i at starting time t, the base year, and G_n^{t+k} is the change rate of the national output in $t + k$ time. Then, the resistance can be expressed as follows.

$$Resistance = \frac{\left(\Delta R_i^{contraction}\right) - \left(\Delta R_i^{contraction}\right)^{expected}}{\left|\left(\Delta R_i^{contraction}\right)^{expected}\right|} \tag{5.2}$$

$\left(\Delta R_i^{contraction}\right)$ is the actual change of economic output of city i during the contraction period.

A resistance that is larger than 0 means that the impact of the COVID-19 crisis on the region is less than that on the national average level, the regional economic resilience level is high, and vice versa.

5.3.3 Geographical Detector Model

Geographical detector model is widely used to identify the effect of different socio-economic factors and its interactions on some particular risks or disease by approach of spatial variance analysis (SVA) (Wang et al., 2010). This model shows comparative advantages of application than the traditional linear regression method when dealing with mixed data in type, as it involves a smaller number of assumptions. Moreover, it takes into account the spatial differentiation characteristics of unit attributes, and can overcome the multicollinearity of variables. Meanwhile, it is more applicable to analyze the influence of the interaction between different variables on the explained variables. More specifically, the basic idea of the method is to measure the degree to which the spatial distribution of regional economic resilience is consistent with that of influencing factors (Wang et al., 2010; Zhou et al., 2018). If there is a significant consistency between a certain factor and regional economic resilience in space, this factor will play a decisive role in the change of regional economic resilience. Since the geographical detector model analyzes the explanatory power of each factor relating to explained variable one by one, other variables are controlled when the explanatory power of a specific factor is calculated. This method thus is applied to explore the geographies of impact factors in shaping regional economic resilience amid COVID-19. The calculation model is as follows:

$$P_{D,U} = 1 - \frac{1}{n\sigma^2{}_U}\sum\nolimits_{i=1}^{m}(n_{D,i} \cdot \sigma^2{}_{U_{D,i}}) \tag{5.3}$$

where $P_{D,U}$ is the power of the influencing factor Di, U is the regional economic resilience, $n_{D,i}$ denotes the number of cities in subareas with the influential factors Di, and $\sigma^2{}_{U_{D,i}}$ denotes the dispersion variance of U in the entire region. $P_{D,U} \in [0,1]$. When $P_{D,U} = 0$, this shows that the regional economic resilience is randomly distributed. The larger $P_{D,U}$ is, the greater the impact of the factors on resilience.

5.3.4 Index Composition and Data Description

Based on our hypotheses and the characteristics of the COVID-19 crisis itself by the time of writing, this paper focuses on four main factors that matter significantly for the economic resilience of Northeastern cities in China. They are basic industrial composition, industrial structure, economic openness, government control factors. The paper chooses GDP per capita and the number of authorized patents as the control factors to reflect the level of regional economic development and innovation. The specific contents and descriptive analysis of relevant indicators are shown in Table 5.1.

First, basic industrial composition, describes the basic characteristic of a region's economic activity, and the stage within the production process. It represents a region's

Table 5.1 Variables and descriptive analysis

Variable	Definition	Unit	Min.	Max.	Mean
Regional economic resilience	The index of resistance		−0.90	1.37	0.07
Secondary industry proportion	The added value of secondary industry as a proportion of GDP	%	11	54	32
Tertiary industry proportion	The added value of tertiary industry as a proportion of GDP	%	37	68	49
Specialization	The index of Krugman specialization		0.32	0.96	0.53
Diversification	The indexes of variety The indexes of unrelated variety		0.76 1.20	1.16 1.56	1.02 1.36
Economic openness	The proportion of total import and export trade in GDP	%	0.38	62.2	11.5
Resumption of production	Rate of major industrial enterprises returned to work	%	72	100	90
Residential community control	Whether to cancel community closure	0 or 1	0	1	0.76
Public transport control	Whether to resume bus operation	0 or 1	0	1	0.76
Economic level	GDP per capita	RMB/ person	21,104	116,948	43,514
Innovation level	The total number of patients	Piece	123	14,532	2196

sketched profile of economic development stage. In this paper, the proportions of secondary and tertiary industry are used as proxies to measure the industrial structure. Second, industrial structure can be well reflected by variables of industrial specialization, variety and unrelated variety. Specialization is expressed by the Krugman specialization index, which can be calculated by the number of employments in various sectors. The calculation formula is as follows:

$$SPEC_i = \sum_{j=1}^{k} \left| V_{i,j} - V_{n,j} \right| \tag{5.4}$$

where $V_{i,j}$ is the proportion of employment in industrial j in total employment in city i, and Vn, J is the proportion of employment in industrial j in total employment in Northeast China.

Industrial diversification here is expressed by overall variety and unrelated variety. Overall variety refers to the degree of association among different industries, including related and unrelated variety, while unrelated variety concerns inter-sectoral diversification without significant complementarity of competences. These two variables are selected because they matter strongly for the resistance to external shocks (Boschma, 2015; Hu & Hassink, 2020). The calculation formula of variety is as follows:

$$V_{i,t}=\sum_{j=1}^{n} E_{i,j}\cdot log(^{1}/_{E_{i,j}}) \qquad E_{i,j}={}^{P_{i,j}}/_{P_i} \tag{5.5}$$

where $P_{i,j}$ is the number of employees in the jth sector of city i at time t, n is the number of all sectors, Pi is the number of employees in city i at time t, Vi, t is the diversity of city i at time t. The calculation formula of unrelated variety is as follows:

$$UV_{i,t}=\sum_{s=1}^{k} P_{i,s}\cdot ln(^{1}/_{P_{i,s}}) \tag{5.6}$$

Pi, s is the proportion of employees in the sth major sector of city i at time t, k is the number of all major industries. According to the differentiated data structure in official statistical yearbooks from 34 Northeastern cities, the number of employees is usually documented based on different sub-sectors. For the comparative setting of the study, we not only draw on data from the statistical yearbooks but also merge the data in various sub-sectors into six major sectors: the primary industry, the secondary industry, the producer service industry, the consumer service industry, the circulation service industry and the social service industry.

Third, a city's economic openness can be measured by the proportion of total import and export trade in GDP. Fourth, compared to the aforementioned three "structural" factors that are significantly associated with vulnerability and resistance, the government control factor is concerned with the role of state agency in resisting and adapting to the crisis. The production resumption rate of major industrial enterprises (annual revenue of 20 million yuan or more) reflects the efforts made by the government to recover a city's economy from the COVID-19 crisis. Variables also include whether a city adopts closed-off policy to its residential communities, and whether a city calls a halt to public transport service. By the end of the first quarter of 2020, the city that cancels community closure policy is set as 0, otherwise it is set as 1; the city that keeps or resumes bus operation is set as 0, otherwise it is set as 1.

Given the limited data availability at the time of writing, our latest data can be only traced back to the end of the first quarter of 2020. More specifically, the first quarter data of the year 2019 and 2020 is used to measure the resistance index, which is derived from the statistical yearbooks of the three provinces in Northeast China and the official website of each city statistics bureau. The data related to "structural" factors including basic industrial composition, industrial structure and economic openness is from the "China City Statistical Yearbook of 2019". All data relating to the government control factor is collected from the official website of each city government and various state-affiliated news sources (electronic newspapers, websites, WeChat platforms etc.).

5.4 Results: Differentiated Resistance to COVID-19

Based on formulas (5.1) and (5.2), the regional economic resilience of Northeast cities midst COVID-19 was measured, in the first quarter of 2020. In Fig. 5.3, the spatial distribution of confirmed cases, as well as it of the resistance value in each city, are presented.

In order to capture the economic resilience characteristics of the study area, we calculate the average resilience value and the total number of confirmed infected cases of the region and its constituent provinces, as shown in Table 5.2.

Figure 7.1 and Table 5.2 show that there is significant spatial heterogeneity in the resistance of Northeast China under COVID-19. Firstly, at the provincial level, the economic resistance of Northeast China is 0.07, which is higher than the national average level. This means that the economy of Northeast China was less hit in the beginning of the crisis, despite its structural disadvantages. With the resistance rate

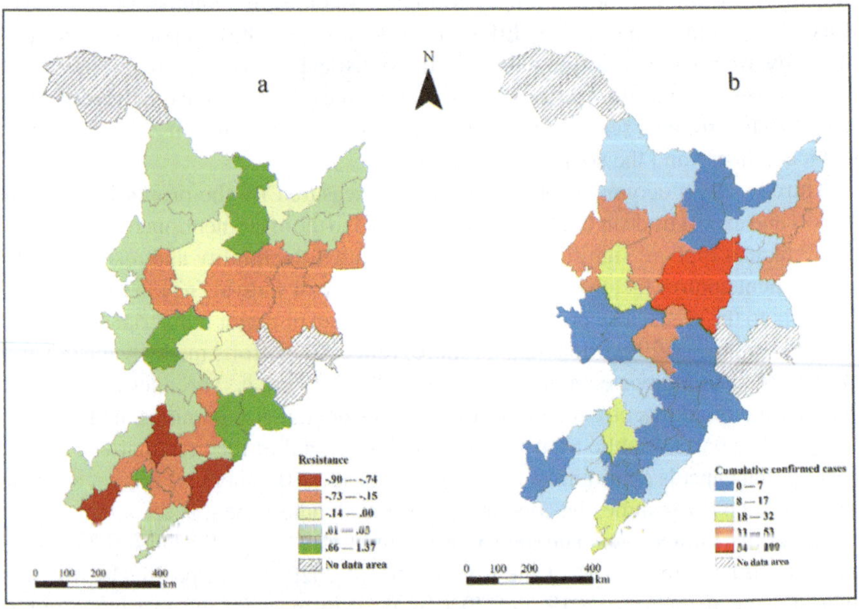

Fig. 5.3 Regional economic resistance and the geographical distribution of the confirmed cases in Northeast cities in China amid COVID-19

Table 5.2 Average value of economic resistance and the total number of confirmed cases

Area	Resistance	Total number of confirmed cases
Liaoning Province	−0.06	140
Jilin Province	0.46	91
Heilongjiang Province	−0.05	481
Northeast China	0.07	712

of 0.46, Jilin performed much better than Liaoning and Heilongjiang to withstand the virus. The resistance is less than 0 in Liaoning and Heilongjiang, indicating that their economies were more severely affected and confronted with a sharp downturn. Second, at the city level, over half of the cities have low resistance (less than 0) and were greatly affected by the shock. The distribution of confirmed cases of COVID-19 is relatively spatially consistent with regional economic resilience. Certainly, cities with more confirmed cases were more severely affected by the shock. Moreover, the cities with weaker resistance show significant spatial agglomeration (Fig. 2.1). This is particular the case for metropolitan cities, such as Shenyang, Changchun-Jilin and Harbin-Mudanjiang. One reason is that compared to small cities, large cities have more global openness and a larger share of secondary and tertiary industry in the economy, they thus were more vulnerable in their economic structure to COVID-19. Moreover, since large cities have more intense human mobility and interactions, the virus is more likely to spreads rapidly and widely. Arguably, in order to contain the pandemic, large cities in Northeast China received stricter restriction measures on human activities, inevitably causing the decline of daily life economies and subsequent weaker resistance.

Our results also indicate that cities with a large number of infected cases generate more negative impact on the economic resistance of their geographically adjunct areas. As such, these neighboring areas to large cities receive more exposure to the virus and thus have to take strict more restrictions measures that significantly hinder regional economic development. Overall, COVID-19 is not an industry-specific shock, but an all-round global crisis with foreseeable long-lasting impact on human activities particularly in megapolitan regions (Adler et al., 2020). Given this, a city's industrial structure alone may play a limited role in affecting regional resilience. Partly in line with hypothesis H4, cities with more global connections and population density are more exposure to COVID-19 with lower resistance.

5.5 Uneven Economic Resistance to COVID-19: What Matters?

To identify the impact factors of regional economic resilience in terms of resistance under COVID-19, the paper classifies the selected independent variables based on ArcGIS Jenks method, and then uses the resistance index as the dependent variable.

Table 5.3 Determinants of regional economic resilience

Variable	q value	P value
Secondary industry share (×1)	−0.535	0.00
Tertiary industry share (×2)	−0.110	0.89
Specialization (×3)	0.313	0.04
Overall variety (×4)	−0.500	0.04
Unrelated variety (×5)	0.182	0.51
Economic openness (×6)	−0.302	0.02
Resumption of production (×7)	0.255	0.33
GDP per capita (×8)	−0.286	0.09
Patient number (×9)	0.357	0.02
Residential community control (×10)	0.077	0.11
Public transport control (×11)	0.020	0.43

We measure the correlation coefficient between the independent and the dependent variable to reflect the action direction between them. Our results are shown in Tables 5.3 and 5.4.

Table 5.3 shows that secondary industry structure, overall variety, innovation ability, specialization, openness and economic development are the main factors shaping the economic resistance of Northeast China to COVID-19. Secondary industry structure (−0.535), tertiary industry structure (−0.110), overall variety (−0.500) and economic openness (−0.302) have negative effects on regional economic resilience. These variables, as all higher for q value, passed the significance test, which indicate that the three variables played an important role in resisting the COVID-19 crisis for Northeastern cities. Since the q values of secondary industry structure and overall variety are the highest, they have the most impact on regional

Table 5.4 The result of interactive detection

	X1	X2	X3	X4	X5	X6	X7	X8	X9	X10	X11
X1	0.535										
X2	0.626	0.11									
X3	0.562	0.663	0.313								
X4	0.733	0.554	0.606	0.5							
X5	0.567	0.293	0.535	0.61	0.182						
X6	0.781	0.625	0.697	0.708	0.469	0.302					
X7	0.626	0.384	0.609	0.694	0.374	0.438	0.255				
X8	0.664	0.394	0.606	0.873	0.351	0.519	0.364	0.286			
X9	0.621	0.768	0.549	0.714	0.568	0.652	0.532	0.548	0.357		
X10	0.538	0.353	0.443	0.567	0.193	0.317	0.277	0.477	0.512	0.077	
X11	0.559	0.258	0.406	0.66	0.286	0.404	0.36	0.27	0.413	NA	0.020

economic resistance. As the results further shown, first, both the secondary and tertiary industry in Northeastern cities were negatively impacted by the pandemic. The tertiary industry, particularly in terms of hospitality, tourism, aviation, sports, finance, education and housing sector, were seriously affected by the crisis and experienced a clear downturn. Cities with a high proportion of the tertiary industry received more negative impact by COVID-19, showing weak and even a lack of resistance. This result thus allows us to confirm H1.

Second, we found that industrial specialization (0.313) shows positive effects on the economic resilience of Northeastern cities under COVID-19. It passed the significance test, and the q value is large. It is surprising that a specialized economic structure is proofed to have stronger regional resistance to the pandemic. This might be explained by two points. For one thing, Northeast China is the country's largest old industrial base, where city economies are mainly state-dominated, resource-based and heavy-industry concentrated (e.g., mining, power, and military). A more specialized economy refers to a higher degree of state domination in the economy, which receive more policy preferences and financial support. For another, state-owned economies in are relatively less globalized: their supply chains and market channels are more domestically oriented and even remain as local, showing stronger resistance to external shocks (Hu & Hassink, 2017b). In other words, a specialized economic structure is not a disadvantage to the COVID-19 crisis, it rather helps to prevent the economy from being hit by the crisis, presenting stronger economic resistance. Meanwhile, our evidence shows that cities with a higher degree of overall variety do not bring about stronger resistance to COVID-19. One key explanation to this is that as long as industries involve labor input COVID-19 clearly impacts them regardless of industry type. This is the case that cities in Northeast China heavily depend on the real economy mainly constituted by labor-intensive manufacturing and resource extraction (Hu and Hassink, 2017c). Furthermore, a more diversified structure in the region may not only refer to specialized diversification (or a higher share of related variety focusing on certain sectors), but also a stronger economy with a larger number of firms and labor. These structural features are rather the weaknesses to prevent COVID-19 from impact. It thus can be argued that cities with a higher degree of overall variety tend to suffer more from the crisis and result in weaker resistance. This evidence then goes against hypothesis H2.

Third, we found that unrelated variety (0.182) have positive effects on the economic resistance of Northeastern cities under COVID-19, despite the q value of unrelated variety is small. This is understandable that cities with more technologically unrelated industry structure may stand for more likelihood of holding industries that are different from traditional manufacturing sectors and thus less affected by COVID-19, such as pharmaceutical, healthcare and IT-related sectors. Therefore, we confirm hypothesis H3 that an unrelated diversified economic structure can function as a shock absorber to the pandemic and can lead to a stronger economic resistance.

Forth, it is also significant that a higher degree of economic openness provides structural disadvantages to withstand the pandemic. This point well echoes with the argument that COVID-19 disrupts manufacturing regions with a strong reliance on global supply chains and trade more severely than others (Bryson & Vanchan, 2020).

Of course, this does not mean that cities with high openness are doomed in a lack of resilience (particularly referring to the recovery dimension) in a long run, but at least they show weaker economic resistance during the early time of COVID-19. Based on that, we are able to confirm H4.

More importantly, government control measures have played a key role in (re)shaping the regional resistance (also recovery) to COVID-19, due to the positive rate of major industrial enterprises that resume production (0.255), residential community control (0.077) and public transport control (0.077). Although the q value of government control measures is relatively small, China's epidemic prevention and control measures have proved that efficient governance plays an important role in the recovery of regional economy. Our results show that city governments in Northeast China took strong control measures, such as factory and business closure, residential community control, traffic control and even city lockdown. These measures were strictly implemented, not only just by law enforcement and but also by people's shared consciousness and reasonability for public safety, and by wider socially constructed trust to the Chinese authority (Chung et al., 2020). On the one hand, for protecting human health and public security, local governments had to sacrifice economic interests and strategically weakened regional resistance. On the other hand, with the pandemic being increasingly contained they also relaxed the restrictions accordingly for economic recovery which was also conducive to regional resistance. The empirical results prove that the return-to-work policy both for public and private sectors, in particular, can improve economic resilience of cities. Our analysis highlight regional resistance is more depending on the role of state agency even at the beginning of the COVID-19 outbreak. It is suggested that although for a long run the existing structural factors in the economy of Northeast China may gradually unfold, regional economic resilience is still very much up to the role of the state. COVID-19 is a very unique crisis, to a large extent, become a test, for governance in which state agency and institutions matter resilience directly.

In the end, we find that the variables of GDP per capita (−0.286) and patents (0.357) play a significant role in promoting the regional economic resilience. The two variables both passed the significance test, with high q value. It shows that high economic level cities were more seriously affected by the crisis. This is because the higher economic level a city is, the higher degree of population concentration the city has and the more rapidly the epidemic spreads. The improvement of regional innovation level can promote the Northeast China to cope with the COVID-19 shock.

Table 4.1 shows the q-values of the interaction detector. These values indicate that the explanatory power (interactive effect) between any two driving factors is always greater than that of a single individual driving factor to regional economic resilience. This also shows that regional economic resilience is a complex and multi-factor involved process, where the interaction between different factors has stronger effect on resilience, and the factors are not independent.

5.6 Conclusions

The notion of regional economic resilience is considered to be highly promising to explain and understand the differences in regional response, adaptability and outcome to shocks (Hu & Hassink, 2020). Two main impact factors, namely structure and agency, have been identified to understand geographically uneven resilience of regions (Bristow & Healy, 2020; Martin & Sunley, 2020). Recent account highlights the role of crisis itself—whose characteristics vary from one to another according to different nature of crises—in affecting regional economic resilience (Martin, 2018). Yet, insufficient work has been done on the topic within a new crisis context. This paper therefore focuses on how regional economies respond to the COVID-19 pandemic—the defining global health crisis of the current time and the greatest challenge human beings have faced since World War Two. By adopting the notion of resistance in regional resilience, it examines how city economies in Northeast China responded to COVID-19 at the early stage of the crisis and explains what matters for regional economic resilience.

Our results show: (1) the economic resistance of Northeast China is higher than it of the national average level, but within the area most cities have weak ability to withstand the pandemic. This implies that a complex set of impact factors matter for regional resistance in Northeast China; (2) regional structural factors still play a role, despite limited in degree. We find that more economic openness, the structure with a high share of secondary and tertiary industry and overall variety, negatively affect the ability of regional resistance; (3) Large city-regions show more vulnerability (less resistance) than small cities, as the former with larger population is more exposed to the virus and thus receives more government restrictions on economic activity (Adler et al., 2020; Ascani et al., 2020). We argue that local state agency plays a role in shaping regional resistance and recovery; (4) An unrelated variety structure can function as a "shock absorber" which disperses the negative impact of the pandemic. However, a specialized economic structure can rather promote regional resistance to COVID-19. This is because: most specialized economies in Northeast China are built up by state-owned enterprises (e.g., in natural resource mining) with domestic supply chains and markets, showing stronger resistance (Hu and Hassink, 2017c).

Conceptually, this paper contributes to the economic resilience literature in the crisis of COVID-19. It is argued that while historically inherited structural configurations—such as industrial composition feature, industrial structure, economic openness do still matter for regional resistance, they function highly differently compared to their roles facing the 2008 financial crisis. The very nature of the crisis, namely as a global pandemic, has led to quite distinct aims of responses, that is, to contain it (at least at the beginning during the crisis) by decreasing human-to-human interactions of any kind for saving lives and public security. We argue that COVID-19 is not a crisis directly toward the economy, it is, at its first place, a political and social crisis that functions as a test for governance at multiple levels to save human lives and sustain socio orders. In this regard, state agency plays a crucial role in shaping economic resistance, through the authoritarian way of planning and re-organizing the economy. Regional resilience under COVID-19 thus is not simply about the vulnerability linking to the attribute of regional economic structure, but rather is more related to the ways in which state agency and institutions interactively response for what purpose and for whom. For instance, local governments can decide when and to what extend economic activities come to a standstill and start to resume and restore. The control measures as we have aforementioned and the quality of governance has a more serious impact on the economy than do the path dependent structural factors (Chung et al., 2020; Ezcurra & Rios, 2019). Certainly, this might be the very case when facing COVID-19 at the early stage, as we shown in the paper.

To sum up, the paper contributes to the existing international literature on COVID-19 by engaging the topic of regional economic resilience. Theoretically, it adds new insights to the regional economic resilience research, namely, regional economic resilience amid COVID-19 and even after the end of it is a mixed process in which actions of vulnerability, resistance, recovery and reorientation may co-exist, which is unlikely to follow a typical sequential order from adaptation to adaptability (Hu & Hassink, 2017a, 2017b; Martin & Sunley, 2020). Moreover, what we find in the paper is that: COVID-19 does not merely refer to negative pressure, but also gives actors leeway to resourcefully try new opportunities (referring to resilience as resourcefulness, see MacKinnon & Derickson, 2013). For instance, some cities do not put production restrictions on all sectors, but may promote some specific sectors such as pharmaceutical, healthcare, video gaming, or other home-based virtual industries (Nicola et al., 2020). Moreover, regional resilience is very much up to the actual developments of COVID-19 well-grounded in the local in term of criticality. In the sense, resilience is about how and to what extent regions enact restriction policies.

Finally, from the policy perspective, future resilience policies under COVID-19 need a stronger awareness of the specific relationship between exposure, vulnerability, resistance and adaptability in the local, and of their tempo-spatial opportunities for positive policy interventions (Bristow & Healy, 2020; Hu & Hassink, 2017a). This requires regional policymakers to take a context-specific view on the nature of COVID-19, namely knowing the actual realities of context shaped by place-specific socio-economic preconditions and the specificities of the COVID-19 impact/process. Moreover, policy actions also need to nurture the resourcefulness of collective agency facing the change of context, and to identify the sequence, priorities, unit of entities and purposes of resilience that are place-based and industrially sensitive/selective (Hu & Yang, 2019). Certainly, our analysis is however not unproblematic, given the limited data availability and our relatively short length of time for research. We think regional responses at the beginning of the pandemic itself might be short-sighted and contextually contingent, with their specific determinants of economic resilience (Gereffi, 2020). For future regional resilience research under COVID-19, more attention needs to be paid to the role of multi-scalar agency, institutions and history relating to place mentality, culture, governance quality and leadership of change, as well as dynamic power relations among various actors and interests in broader geographical political economy in a long run.

Acknowledgements Liangang Li from Shandong Normal University and Ke Dong from Zhejiang Normal University have made great contributions to this chapter. We express our heartfelt thanks to them.

References

Adler, P., Florida, R., & Hartt, M. (2020). Mega regions and pandemics. *Tijdschrift Voor Economiche En Sociale Geograhie, 111*(3), 465–481.

Angulo, A. M., Mur, J., & Trívez, F. J. (2018). Measuring resilience to economic shocks: An application to Spain. *The Annals of Regional Science, 60*(2), 349–373.

Ascani, A., Faggian, A., & Montresor, S. (2020). The geography of COVID-19 and the structure of local economies: the case of Italy. Working Papers No. 1. Gran Sasso Science Institute.

Aslund, A. (2020). Responses to the COVID-19 crisis in Russia, Ukraine, and Belarus. *Eurasian Geography and Economics*, 1–14.

Asongu, S. A., Diop, S., & Nnanna, J. (2020). The geography of the effectiveness and consequences of Covid-19 measures: Global evidence. *Journal of Public Affairs*, 1–9.

Atikian, J. (2013) What is industrial structure? In *Industrial shift: The structure of the new world economy*. Palgrave Pivot.

Bailey, D., Clark, J., Colombelli, A., et al. (2020). Regions in a time of pandemic. *Regional Studies, 54*(9), 1163–1174.

Balland, P. A., Rigby, D. L., & Boschma, R. (2015). The technological resilience of US Cities. *Cambridge Journal of Regions, Economy and Society, 8*(2), 167–184.

Boschma, R. (2015). Towards an evolutionary perspective on regional resilience. *Regional Studies, 49*(5), 733–751.

Bristow, G., & Healy, A. (2014). Regional resilience: An agency perspective. *Regional Studies, 48*(5), 923–935.

Bristow, G., & Healy, A. (2020). Introduction to the handbook on regional economic Resilience. In G. Bristow & A. Healy (Eds.), *Handbook on regional economic resilience* (pp. 1–8). Edward Elgar.

Bryson, J. R. & Vanchan, V. (2020). COVID-19 and alternative conceptualizations of value and risk in GPN research. *ijdschrift voor Economiche en Sociale Geograhie, 111*(3), 530–542.

Chan, K. W., Gentile, M., Kinossian, N., Oakes, T., & Young, C. (2020). "More-than-viral" Eurasian geographies of the covid-19 pandemic: interconnections, inequalities, and geopolitics. *Eurasian Geography and Economics*, 1–20.

Chen, S. Q., & Bonanno, G. A. (2020). Psychological adjustment during the global outbreak of COVID-19: A resilience perspective. *Psychological Trauma: Theory, Research, Practice, and Policy, 12*(S1), S51.

Chen, S. M., Yang, J. T., Yang, W., Wang, C., & Bärnighausen, T. (2020). COVID-19 control in China during mass population movements at New Year. *The Lancet, 10266*(395), 764–766.

Chung, C. K. L., Xu, J., & Zhang, M. (2020). Geographies of Covid-19: how space and virus shape each other. *Asian Geographer*, 1–18.

Content, J., & Frenken, K. (2016). Related variety and economic development: A literature review. *European Planning Studies, 24*(12), 2097–2112.

David, L. (2018). Agency and resilience in the time of regional economic crisis. *European Planning Studies, 26*(5), 1041–1059.

Dodds, K., Broto, V. C., Detterbeck, K., et al. (2020). The COVID-19 pandemic: Territorial, political and governance dimensions of the crisis. *Territory, Politics, Governance, 8*(3), 289–298.

Eraydin, A. (2016). Attributes and characteristics of regional resilience: Defining and measuring the resilience of Turkish regions. *Regional Studies, 50*(4), 600–614.

Ezcurra, R., & Rios, V. (2019). Quality of government and regional resilience in the European Union. Evidence from the Great Recession. *Papers in Regional Science, 98*(3), 1267–1290.
Foster, K. A. (2007). A case study approach to understanding regional resilience. Working Paper Number 2007–08. Institution of Urban and Regional Development, Berkeley, CA.
Garavaglia, C., Sancino, A., & Trivellato, B. (2020). Italian mayors and the management of COVID-19 adaptative leadership for organizing local governance. *Eurasian Geography and Economics*. https://doi.org/10.1080/15387216.2020.1845222
Gereffi, G. (2020). What does the COVID-19 pandemic teach us about global value chains? The case of medical supplies. *Journal of International Business Policy, 3*(3), 287–301.
Giannakis, E., & Bruggeman, A. (2017a). Economic crisis and regional resilience: Evidence from Greece. *Papers in Regional Science, 96*(3), 451–476.
Giannakis, E., & Bruggeman, A. (2017b). Determinants of regional resilience to economic crisis: A European perspective. *European Planning Studies, 25*(8), 1394–1415.
Grabher, G., & Stark, D. (1997). Organizing diversity: Evolutionary theory, network analysis and postsocialism. *Regional Studies, 31*, 533–544.
Grundy-Warr, C., & Lin, S. (2020). COVID-19 Geopolitics: Silence and Erasure in Cambodia and Myanmar in Times of Pandemic. *Eurasian Geography and Economics, 61*, 4–5.
Hassink, R. (2010). Regional resilience: A promising concept to explain differences in regional economic adaptability? *Cambridge Journal of Regions, Economy and Society, 3*(1), 45–58.
He, A. J., Shi, Y., & Liu, H. (2020). Crisis governance, Chinese style: Distinctive features of China's responses to the Govid-19 pandemic. *Policy Design and Practice, 3*(3), 242–258.
Hill, E. W., Wial, H., & Wolman, H. (2008). Exploring regional resilience. Working Paper Number 2008–04. Institute of Urban and Regional Development.
Hu, X. (2015). *Exploring differentiated economic adaptation and adaptability of old industrial areas in transitional China*. Kiel University Press.
Hu, X., & Hassink, R. (2017a). Exploring adaptation and adaptability in uneven economic resilience: A tale of two Chinese mining regions. *Cambridge Journal of Regions, Economy and Society, 10*(3), 527–541.
Hu, X., & Hassink, R. (2017b). Place leadership with Chinese characteristics? A case study of the Zaozhuang coal-mining region in transition. *Regional Studies, 51*(2), 224–234.
Hu, X., & Hassink, R. (2020). Adaptation, adaptability and regional economic resilience: A conceptual framework. In G. Bristow & A. Healy (Eds.), *Handbook on regional economic resilience* (pp. 54–68). Edward Elgar.
Hu, X., & Yang, C. (2019). Institutional change and divergent economic resilience: Path development of two resource-depleted cities in China. *Urban Studies, 56*(16), 3466–3485.
Hu, X., & Hassink, R. (2017c). New perspectives on restructuring of old industrial areas in China: A critical review and research agenda. *Chinese Geographical Science, 27*(1), 110–122.
Ivanov, D., & Dolgui, A. (2020). Viability of intertwined supply networks: extending the supply chain resilience angles towards survivability. A position paper motivated by COVID-19 outbreak. *International Journal of Production Research, 58*(10), 2904–2915.
JHU (Johns Hopkins University). (2021). COVID-19 Dashboard by the Center for Systems Science and Engineering (CSSE) at Johns Hopkins University (JHU). ArcGIS. Johns Hopkins University. Retrieved 13 May 2021.
Kakderi, C., & Tasopoulou, A. (2017). Regional economic resilience: The role of national and regional policies. *European Planning Studies, 25*(8), 1435–1453.
Klassen, S., & Murphy, S. (2020). Equity as both a means and an end: Lessons for resilient food systems from COVID-19. *World Development, 136*, 105104.
Lawreniuk, S. (2020). Necrocapitalist networks: COVID-19 and the 'dark side' of economic geography. *Dialogues in Human Geography, 10*(2), 199–202.
Li, L. G., Zhang, P. Y., & Li, X. (2019). Regional economic resilience of the old industrial bases in China: A case study of Liaoning Province. *Sustainability, 11*(3), 723.
Liu, W. D. (2020). The impacts of COVID-19 pandemic on the development of economic globalization. *Geographical Research, 39*(7), 1439–1449. (In Chinese).

MacKinnon, D., & Derickson, K. D. (2013). From resilience to resourcefulness: A critique of resilience policy and activism. *Progress in Human Geography, 37*(2), 253–270.

Martin, R. (2012). Regional economic resilience, hysteresis and recessionary shocks. *Journal of Economic Geography, 12*(1), 1–32.

Martin, R., & Sunley, P. (2015). On the notion of regional economic resilience: Conceptualization and explanation. *Journal of Economic Geography, 15*, 1–42.

Martin, R., & Sunley, P. (2020). Regional economic resilience: Evolution and evaluation. In G. Bristow & A. Healy (Eds.), *Handbook on regional economic resilience* (pp. 10–35). Edward Elgar.

Martin, R., Sunley, P., Gardiner, B., & Tyler, P. (2016). How regions react to recessions: Resilience and the role of economic structure. *Regional Studies, 50*(4), 561–585.

Martin, R. (2018). Shocking aspects of regional development: Towards an economic geography of resilience. In: G. Clark, M. Gertler, M. P. Feldman & D. W. Jcik (Eds.), *The new Oxford handbook of economic geography* (pp. 839–864). Oxford University Press.

NBSC (National Bureau of Statistics of China). (2020). National Economy Withstood the Impact of COVID-19 in the First Two Months. Retrieved from: www.stats.gov.cn/english/PressRelease/202003/t20200316_1732244.html

Nicola, M., Alsafi, Z., Sohrabi, C., et al. (2020). The socio-economic implications of the coronavirus pandemic (COVID-19): A review. *International Journal of Surgery, 78*, 185–193.

OECD. (2020). A systemic resilience approach to dealing with Covid-19 and future shocks. In *OECD policy responses to coronavirus (COVID-19)*. OECD Publishing.

Pike, A., Dawley, S., & Tomaney, J. (2010). Resilience, adaptation and adaptability. *Cambridge Journal of Regions, Economy and Society, 3*(1), 59–70.

Ringwood, L., Watson, P., & Lewin, P. (2019). A quantitative method for measuring regional economic resilience to the great recession. *Growth and Change, 50*(1), 381–402.

Rose-Redwood, R., Kitchin, R., Apostolopoulou, E., et al. (2020). Geographies of the COVID-19 pandemic. *Dialogues in Human Geography, 10*(2), 97–106.

Simmie, J., & Martin, R. (2010). The economic resilience of regions: Towards an evolutionary approach. *Cambridge Journal of Regions, Economy and Society., 3*(1), 27–43.

Tan, J. T., Lo, K., Qiu, F. D., Zhang, X. L., & Zhao, H. B. (2020b). Regional economic resilience of resource-based cities and influential factors during economic crises in China. *Growth and Change, 51*(1), 362–381.

Tan, J. T., Zhang, P. Y., Lo, K., Li, J., & Liu, S. W. (2017). Conceptualizing and measuring economic resilience of resource-based cities: Case study of Northeast China. *Chinese Geographical Science, 27*, 471–481.

Tan, J. T. Hu, X., Hassink, R. & Ni, J. W. (2020a). Industrial structure or agency: what affects regional economic resilience? Evidence from resource-based cities in China. Cities. https://doi.org/10.1016/j.cities.2020.102906

Wang, J., Du, D., Ye, W., & Yang, H. (2020). The development of COVID-19 in China: Spatial diffusion and geographical pattern. *Geographical Research, 39*(7), 1450–1462. (In Chinese).

Wang, J. F., Li, X. H., Christakos, G., et al. (2010). Geographical detectors-based health risk assessment and its application in the neural tube defects study of the Heshun Region, China. *International Journal of Geographical Information Science, 24*(1), 107–127.

Welsh, M. (2014). Resilience and responsibility: Governing uncertainty in a complex world. *The Geographical Journal, 180*(1), 15–26.

Zhou, C., Chen, J., & Wang, S. (2018). Examining the effects of socioeconomic development on fine particulate matter (PM2. 5) in China's cities using spatial regression and the geographical detector technique. *Science of the Total Environment, 619*, 436–445.

Zhu, S., He, C., & Zhou, Y. (2017). How to jump further and catch up? Path-breaking in an uneven industrial space. *Journal of Economic Geography, 17*(3), 521–545.

Chapter 6
Exploring Adaptation and Adaptability in Uneven Economic Resilience: A Tale of Two Chinese Mining Regions

The wise adapt themselves to circumstances, as water moulds itself to the pitcher.
——Chinese Proverb

6.1 Introduction

In recent years, the notion of 'regional resilience' has become a buzzword in economic geography and regional studies (Martin & Sunley, 2015). For one thing, the project strongly reflects the ongoing 'recovery' policy agenda at multiple spatial levels after the 2008 global economic crisis (Foster, 2007). For another, the notion's emphasis on history-related mechanisms of regional dynamics closely resonates with the 'evolutionary turn' in economic geography (Boschma, 2015; Hassink, 2010). By rejecting the conventional (multi-) equilibrium view, the notion basically offers a promising intellectual tool to understand how a regional economy responds to and recovers from shocks or disturbances (Martin, 2012). However, the understanding of the notion is not unproblematic. Many studies consider resilience as a metaphor, namely a reductionist analogy between the workings of nature and those of man-made regional economic systems, rather than a thorough concept (Pendall et al., 2010). Moreover, in many cases, the term 'resilience' is used in a loose way, either as a fashionable policy instrument, or as a wishful thought for future development (Martin & Sunley, 2015). This has raised worries that regional resilience might become a 'fuzzy concept' akin to 'clusters' (Christopherson et al., 2010, p. 4; Gong & Hassink, 2017), losing its conceptual value and durability in future research.

Leading economic geographers believe that regional resilience should not be a transient notion, despite its variegated conceptual roots (Martin, 2012; Martin &

(This chapter is a reprint version from: Hu, X., & Hassink, R. (2017). Exploring adaptation and adaptability in uneven economic resilience: A tale of two Chinese mining regions. Cambridge Journal of Regions, Economy and Society, 10(3), 527–541.)

X. Hu, *Uneven Economic Resilience of Old Industrial Cities in China*,
Urban Sustainability, https://doi.org/10.1007/978-981-99-9279-9_6

Sunley, 2015). Christopherson et al., (2010, p. 3) point out that resilience echoes, in a fundamental sense, 'a very old and enduring question' in economic geography, that is, why some regional economies are capable of overcoming all sorts of shocks while others are not (Hassink, 2010; Simmie & Martin, 2010). Furthermore, as suggested by Martin and Sunley (2015, p. 3), 'regional resilience is part of the process of geographically uneven economic evolution and development, and should be theorised and analysed as such'. Following this, a recent body of literature shows that the notion of regional resilience can be better conceptually valued in a *long-term* regional uneven development framework (Cambridge Journal of Regions, Economy and Society, 2015; Regional Studies, 2016). The aim of this article is to contribute to this research trend, through a detailed comparative study on industrial dynamics in two coal-mining regions in China linking adaptation and adaptability—two often-adopted but not clearly clarified notions in the literature.

Informed by the complex adaptive systems theory, 'adaptation' and 'adaptability' are the most debated notions in urban and regional resilience research (Boschma, 2015; Hassink, 2010; Martin, 2012; Pike et al., 2010). In many cases, adaptation is portrayed as a path-dependent process maintaining existing development paths or primary functionalities of a system (Adger, 2010). It often involves short-term quantitative, non-radical changes or adjustments of the regional agenda, strategy and action, in order to regain or maintain the economy's 'core performances' (Martin et al., 2016), while adaptability is believed to be a more 'transformation-over-stabilisation' term. The latter often refers to a long-term adaptive ability of an economic system seeking qualitative changes or shifting of industrial paths, developmental modes and systematic institutions (Boschma, 2015; Chapman et al., 2004; Pike et al., 2010). These assumptions, despite good reasons to support (Adger, 2010; Grabher, 1995; Grabher & Stark, 1997), lack rigour, which might undermine the conceptualisation and operationalisation of regional resilience.

Adaptation and adaptability are somehow preconditioned with competing meanings, but they are by no means separated. A number of studies indicate that the tensions between the two are dynamic, and that should be at the core when exploring the geographies of resilience. Boschma (2015, p. 734) argues that resilience is about 'how to secure adaptation and adaptability simultaneously', and related variety is the key to strike such an industrial/institutional balance to make a region resilient in a long-run context. Moreover, according to Ibert and Schmidt (2014), adaptation and adaptability are integral dimensions of resilience, and they can be coherently connected when, for instance, strategic actors enact 'purposive adaptation' that nurtures adaptability for regional industrial path shifting and upgrading (Tomlinson & Branston, 2014). As has been emphasised by Pike et al. (2010), MacKinnon and Derickson (2013) in a Western context, the relationship between adaptation and adaptability is complex and dynamic, which is not only shaped by firm agency and embedded in a regional context, but also mediated by the interplay between the state, capital, labour and society in a broader multi-scalar institutional environment. The under-explored question here is not about whether adaptation and adaptability are related, but about how they interact, and how possibly different forms and extents of interaction could lead to uneven regional resilience.

This article therefore seeks to contribute to this debate by developing a conceptual approach concerning adaptation and adaptability in order to advance the understanding of regional economic resilience in a long-run context. We suggest an approach of adaptation and adaptability in an evolutionary and dialectical way. This approach will be applied to empirical case studies of two coal-mining municipal regions in China, namely, Zaozhuang in Shandong province and Fuxin in Liaoning province, from which we expect interesting insights in the abovementioned interplay between the state, capital, labour and society in a different non-Western multi-scalar institutional environment.

The article is organised as follows: based on the above-mentioned critique on the trade-off account on adaptation and adaptability, the second section proposes a typology of their relationship relating to regional resilience. After introducing the research context and methodology in the third section, the following section analyses comparatively how and to what extent adaptation (the old mining path) and adaptability (newly emerged industrial paths) interact in the case regions since 2000, and explains why Zaozhuang is more resilient than Fuxin. Finally, the last section summarises the article, and concludes by calling for a need to investigate 'quality' and 'process characteristics' of long-term regional resilience.

6.2 Placing Adaptation and Adaptability: A Typology of the Relationship

Long-term resilience in general embodies two interdependent and co-evolving dimensions, namely, resilience process and resilience capability. Adaptation can be seen as resilience process, and adaptability can be related to resilience capability.

The conventional idea of adaptation is preconditioned by a short-term post-crisis context. This discourse might be inappropriate in a long-term setting. A crisis (even a one-off shock) has a long-lasting impact in nature known as 'remanence' (Cross et al., 2009), which means adaptation may never reach an end. But there are also are multi-scalar heterogeneous actors involved in adaptation, some of whom may not merely respond for stability, but may consciously develop a far-reaching agenda for structural change and/or long-term prosperity. Furthermore, history is an important but not deterministic input for adaptation being more or less path-dependent. The ways in which adaptation unfolds are open-ended, contingent to evolving tempo-spatial contexts. They can lead to divergent resilience processes and outcomes, such as 'resistance', 'recovery', 'renewal' and 'reorientation', as suggested by Martin (2012).

Adaptability, by its literal meaning, can be interpreted as an adaptive ability. In the short run, the notion is likely to resemble 'recoverability' from shocks (Martin & Sunley, 2015). But, in the long run it represents a constructive capacity, or a potential, seeking to make a regional economy (more) resilient. Long-term adaptability therefore is more likely about an ability to change and transform, associated with 'resourcefulness' (MacKinnon & Derickson, 2013) and 'robustness' of resilience (Martin & Sunley, 2015). Moreover, adaptability does not emerge from scratch. It is rather, to a certain extent, bred out of, and nurtured by, adaptation processes in which entrepreneurial agents search a novel way of doing things in order to fit into changing and uncertain contexts (Sensier et al., 2016).

Several studies have shown that adaptation and adaptability are more likely to co-exist nd interact in a dynamic way (Chapman et al., 2004; Ibert & Schmidt, 2014; Morgan, 2013). This can be well exemplified by the coexistence of existing old industries (as a main object of *adaptatio*n) and newly-emerging industries (as an *indicator of adaptability*) in a region (Boschma, 2015; Pike et al., 2010). We thus suggest that the ways in which established old industries (refer to *adaptation*) and newly-forming ones (refer to *adaptability*) interact are place dependent, and can be arguably indicative of uneven economic resilience. To be clear, we classify (at least) three general types of adaptation—adaptability relationships, namely, the separated, competing and reciprocal relationships. We also provide several examples of the types (Table 6.1).

Type A means that regions achieve adaptability at the cost of adaptation, and such adaptability itself is compromised by adaptation (Bathelt et al., 2013). This type can be observed in regions where traditional and new industries co-exist and compete with each other. Many old industrial cities (OICs), for instance, are keen to introduce exogenous knowledge, technologies and labour for path creation as compensation to the loss of old industries (Dawley et al., 2015; Isaksen, 2015). But it is difficult for

Table 6.1 Varieties of adaptation–adaptability relationships in the long run setting

Types of relationship	Six Key examples
Type A: Competing (harmful resilience)	A1. New industries are transplanted and manipulated by, and primarily beneficial for extra-local (national, international) actors, becoming competitors to local ones A2. Creation of newness by fast policy as transient responses to contextual changes with a neglect of local interests
Type B: Separated (weak resilience)	B1. Entirely new industries are implanted with few linkages with regional milieu, like 'cathedrals in the desert' B2. Actors adapt by protecting existing paths and simultaneously developing new industries unrelated to existing ones
Type C: Reciprocal (strong resilience)	C1. Endogenous purposive adaptation in relation to knowledge creation, spin-off, technological change, institutional restructuring can lead to new industry creation and/or renewal C2. Inflow of exogenous (national, international) sources that match with local assets; Actors get actively involved in a resource anchoring and social embedding process

exogenous sources to permeate local established industrial niches, given the lack of social learning (Coenen et al., 2014). Regional economies can also become 'locked in' to global production networks (GPNs) controlled by powerful transnational corporations (MacKinnon, 2012). These regions' adaptation may become internally divergent, in which actors enact with incoherent interests and powers (Dawley et al., 2010). Such competing relationship between adaptation and adaptability may also exist in regions (as Example A2 in Table 6.1) where local policy-makers frequently shift development strategies/policies to fit to changing contexts, but with limited concerns about existing industries (Peck & Theodore, 2015). Although in this context, new path formation might occur in regional economies, it is often shortsighted, hardly enabling system change. Thus, Type A generates harmful resilience, which undermines long-term regional development.

Type B refers to relatively weak resilience, by which adaptation and adaptability interact to a limited extent, and even separately co-exist in a region. In practice, multi-faced strategies may be pursued in regions. That is, actors maintain or protect old industries and simultaneously develop entirely new structures (Morgan, 2013). This can be achieved by 'transplantation' of new sectors from elsewhere (Martin & Sunley, 2006), or by an endogenous strategy to foster a variety unrelated to existing industries and technologies (Frenken et al., 2007). In less-favoured regions, exogenously implanted industries are often vertically integrated into extra-local firms and networks, so as to avoid tensions with vested interests of old industrial systems (Hansen, 1992). As such, adaptability might be enhanced by an increase in industrial/institutional heterogeneity, but adaptation (the core institutions and their main functions) stays virtually unchanged. The separation may result in a highly incoherent institutional milieu, preventing actors from aligning and combining resources for building up industrial specialisations. A lack of industrial focus would potentially lead to undermine regional advantages and weaken long-term resilience.

Type C represents an interactive and intertwined relationship between adaptation and adaptability. In some conditions adaptation and adaptability can be harmonised in a reciprocal manner, which is conducive to long-term regional resilience (Ibert & Schmidt, 2014; Tomlinson & Branston, 2014). Science-based innovation, spin-offs, and entrepreneurship can be seen as one example (Example C1 in Table 6.1) that can lead to positive adaptability and subsequently can reshape the previous form of regional adaptation (Feldman, 2007). In another case, some regional actors are aware of how to exploit global opportunities by strategically making existing regional assets 'plugged' into GPNs, leading to regional value creation and industrial upgrading (MacKinnon, 2012). This is a typical example manifesting that adaptation can enable adaptability through the role of strategic agency, and adaptability can in turn give feedback to adaptation resulting in path renewal, for instance.

Additionally, some regions, particularly 'institutionally thin' peripheries, are not necessarily related to lock-ins and weak resilience. By analysing the formation of the Örnsköldsvik bio-refinery industry in Sweden, Arbuthnott et al. (2010) highlight the importance of social interaction processes between traditional and new industry actors for reframing complementary business concepts in the region that helped to engender industrial renewal. Adaptability, albeit exogenously sourced, can be nurtured locally, in order to be more place-embedded, generating long-lasting positive impacts. It is evidenced that the state can act as a mediator actively involving the recombination of local and non-local knowledge, through policy interventions (Dawley et al., 2015; Isaksen & Trippl, 2016). As such, adaptability benefits from the existing adaptation process and simultaneously pumps transformative stimuli into the existing process of adaptation. Such type involves a co-evolutionary and dialectical process towards institutionally coherent and industrially complementary regional restructuring. This type can potentially form strong resilience.

As we have shown above, the adaptation–adaptability relationship is shaped by mixed factors, including economic-structural and political-institutional ones, characterised by endogenous and exogenous dynamics. These factors do not necessarily directly affect adaptation and adaptability, but always influence the actions and practices of local and extra-local actors (Robinson & Carson, 2016). Of all impact factors, some are the most critical ones that can potentially answer the question of 'adaptation to what' or 'adaptation for what' (Smit et al., 2000). They play a key role in making a region's adaptation–adaptability relationship specific. Regional resilience is, from an agency perspective, strongly linked to the interests of local actors (for example, entrepreneurs, government officials and stakeholders) (Bristow & Healy, 2014). But more importantly, it may also relate to the ways in which local actors interact with, and are influenced by, extra-local, higher and broader levels of power and institutions. For instance, the spatial heterogeneities of local-central politics are common in China, but have not been well examined yet (Huang, 1996). Such heterogeneities are well reflected by the Chinese government convention of local responses to national policies, namely 'the top has policies, while the locality has countermeasures' (shangyou zhengce, xiayou duice, 上有政策, 下有对策) (Economist, 2015). A certain form of local-central relationship in a region would basically set a region's scope and capability of development for the long term and can therefore be seen as a core determinant of resilience (Chien, 2007). Our conceptual framework, therefore, highlights the role of the spatially differentiated agency and its varied impact factors in constituting varieties of adaptation –adaptability relationships explaining uneven resilience.

The above-mentioned varieties of relationships between adaptation and adaptability provide a framework to analyse uneven resilience of regions. In the next sections, we will explore uneven resilience through empirical studies of old industry evolution, new industry creation, and their relationship in two coal-mining regions in China. The Chinese transitional and emerging context, where dynamic heterogeneity and spatiality of agencies and institutions, and broadly, variegated of capitalism and developmental paths/modes co-exist within and among regions, provides a promising ground for studying uneven regional resilience (Zhang & Peck, 2016). The empirical investigations focus on three main questions: what are the characteristics of the mining industry evolution and the emergence of new industries in the two regions? How and to what extent are the old industry and new industries related to each other? According to the typology, what types of adaptation—adaptability relationships can be found in the regions?

6.3 Context and Methodology

Two coal-mining regions in China, namely Zaozhuang in Shandong province and Fuxin in Liaoning province (Fig. 6.1), were selected as case studies for three reasons. First, both regional economies have been denominated by state-owned enterprises (SOEs) specialising in coal mining for decades. Moreover, in 2000, in both regions a 'slow burn' crisis of coal depletion started, for which both were designated as 'Resource-exhausted Regions' by the central state (State Council, 2013). They therefore share a similar adaptation agenda of coping with this chronic crisis. Second, adaptability can be observed in both regions. Zaozhuang recently has diversified its economy by boosting the coal-chemical and tourism industry (Hu, 2017). Fuxin seems to have changed the coal-based mono-structure by embarking on several new industrial paths, such as agriculture/food (Wang et al., 2014), and wind power (Hu, 2014). Third, despite the similarities, significant regional economic gaps remain. Figure 6.2 shows how the disparities in the employment capacity of the two regions have increased since 1985. In that year, the regions had nearly identical employment, but since then Zaozhuang presented a much stronger adaptive momentum than Fuxin. In particular, the Zaozhuang economy has been growing rapidly since 2000, and by 2013 its employment more than doubled compared to Fuxin. The latter, however, seems trapped into a low-speed and fluctuating development mode, lacking robust dynamism.

Based on the conceptual approach and the context described above, the empirical part here aims to comparatively examine the interplays between the coal mining industry (related to adaptation) and newly emerging industries (related to adaptability) in the regions since 2000. The tourism industry in Zaozhuang, the wind

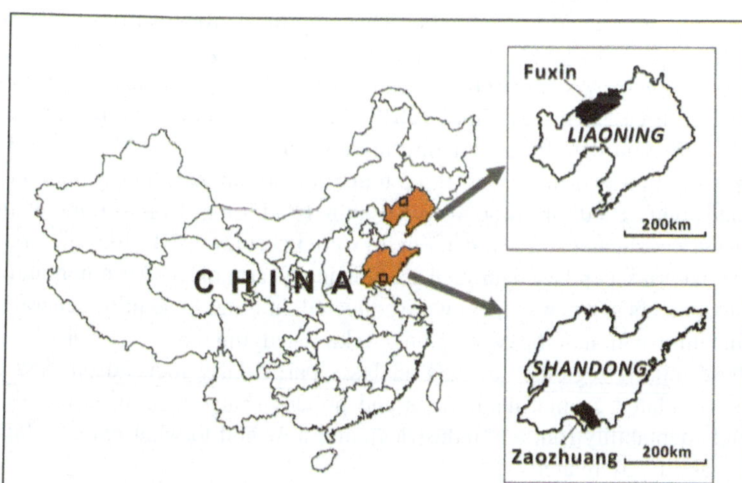

Fig. 6.1 The geographical location of Zaozhuang and Fuxin

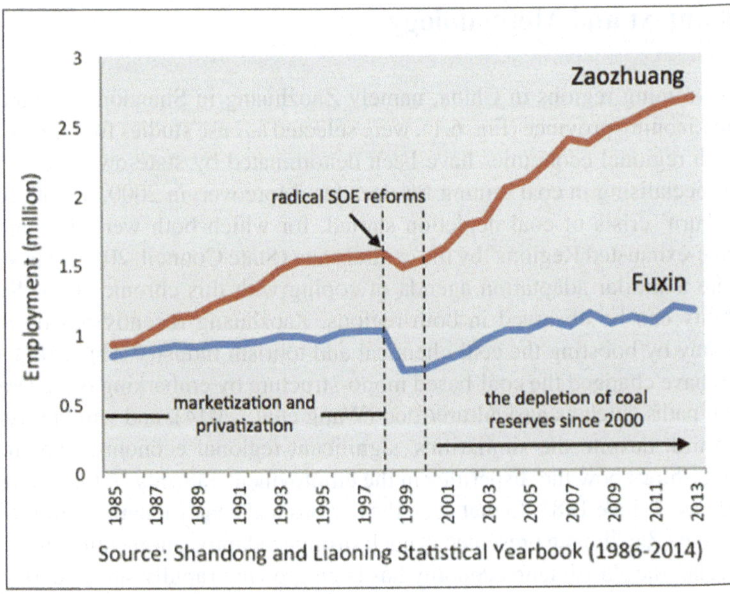

Fig. 6.2 Employment developmental paths in Fuxin and Zaozhuang, 1985–2013

power industry and the agriculture industry in Fuxin, were selected as main research objects and focus. These selected industries can represent the major emerging industrial paths of the regions since 2000, alongside the prevailing mining path. In this study, the restructuring of the coal-mining regions is analysed from a perspective of adaptation and adaptability on the selected new industries and the coal industry and how they are related to each other. Our empirical investigations are based on two on-site fieldworks involving indepth interviews in 2013 and 2014, with a time length of 5 months. As research assistants in the 'National Transformation Planning on Resource-exhausted Cities (2013–2020)', the authors were able to gain access to sufficient data by conducting in-depth interviews wit 13 coal SOEs (11 in Zaozhuang and two in Fuxin), three tourism firms in Zaozhuang, and 11 wind power and eight agriculture/food firms in Fuxin. In addition, 16 government authorities (nine in Zaozhuang and seven in Fuxin) were interviewed. Besides the first-hand data, a variety of secondary data such as statistical yearbooks, governmental reports, policy documents and newspapers were collected, in order to acquire general socio-economic information of the two regions. Acknowledging the strong role of the state in Chinese mining regions, our data analyses particularly focused on how multi-scalar state-related agents, institutions and polices shape adaptation (on the coal path) and adaptability (on new industries), and how and to what extent adaptation and adaptability are related.

6.4 Empirical Analysis

6.4.1 Zaozhuang

Given the abundance of coal reserves, and being the birthplace of China's first domestic coal company in 1878, Zaozhuang municipality has been labelled as *meicheng* (coal city, 煤城). Although the coal industry significantly contributed to local economic growth over decades, it experienced a long-lasting low-speed development during the market reforms in the 1980 and 1990s. In 1998, the coal-based economy, monopolised by Zaozhuang Coal Mine (ZCM), a provincial SOE, was severely hit by a top–down radical ownership reform. This reform cut off 90% of service-based collective-owned subsidiaries and shut down 30% of coal production-based state-owned plants of ZCM, resulting in massive job losses and thus an economic downturn. In response, local officials set up a remedial agenda, giving priority to securing employment and local social stability. As a result, eight municipal/county-level coal SOEs were established by the municipal government between 1999 and 2000, to recruit laid-offs from ZCM. This strategy led to a quick recovery of the local economy from the 1998 reform (Fig. 6.2). Indeed, the strategy played an essential role in helping the municipal state to reap more power (partially from ZCM) to intervene in the local economy. As several interviewees commonly put, by so doing, namely 'raising our own children [running locally-controlled coal businesses]', the municipal state became more autonomous, both in terms of public finance and administration.

Despite the reform, ZCM's ownership still remained provincial, and still possessed plenty of coal resources, accounting for 70% of the total coal reserves in the region. Having only 30% of the coal reserves, the local SOEs were soon confronted with the problem of coal depletion within a few years. The impossibilities of getting access to coal resources controlled by the provincial ZCM forced local (municipal) elites to think of new options to survive. However, this purposive adaptation was timely eroded by a set of top-down national policies deregulating cross-regional mergers and acquisitions (M&As) of coal SOEs from 2005 onwards (Hu, 2017). These policies led to a new round of adaptation characterised by path extension. As the general manager of one local coal SOE put it:

> The national policy temporarily eliminated our anxiety about the coal exhaustion, but it also, to a large extent, demotivated coal entrepreneurs to try new things. With the increase of coal prices between 2005 and 2008, there was simply no reason for us to do something else. We only focused on the old business which, at that time, was extremely profitable. (Interview, translated from Chinese)

However, such a mode of adaptation was challenged when it came to a slump in the coal market affected by the 2008 global crisis. Against that backdrop, a new mayor was appointed by the provincial government who was expected to be able to cope with Zaozhuang's problems and challenges. Unlike many other city mayors in China, who are conventional in economic policy-making and governance, the new mayor viewed the existing adaptation mode as unsustainable and hence saw a need to change it. By using formal and informal tactics and power, the mayor managed to align local state elites and coal industry stakeholders to transform 'the old way of thinking', collectively forming a consensus to foster coal-unrelated, ideally low-carbon, sectors (Interview with a senior official in the municipal state). Moreover, a state-led industrial strategy, namely Cultivating Cultural-Tourism Industries, was launched in 2008. Based on a combination of favourable local conditions and advantageous national policy directions, it was also designed in an attempt to diversify the over-reliant coal economy. As one senior official in the Municipal State Office put it:

> Tourism, in fact, contributes less to GDP, compared to manufacturing, but it can increase employment, and also can improve regional images for attracting inward investment. More importantly, developing tourism, at that time, fits very well to, and can benefit from the national policies supporting 'Red Tourism'[1] (Interview, translated from Chinese)

However, the local state lacked endogenous funds to implement the new strategy. Local officials were merely able to draw resources from locally owned (municipal and county-level) coal SOEs. To put the strategy into effect, the mayor exercised coercive power to engage local coal SOEs into the tourism industry, but, simultaneously, fully supported them to carry out cross-regional M&As. By doing that, the new industrial formation did not harm the old mode of adaptation. Consequently, in 2009 a large tourism theme park—the 'Taierzhuang Ancien Town' (台儿庄古镇) was established. Every local coal SOE was co-opted to run tourism-related businesses, such as tourism-related service firms, hotels and museums (Interviews with several local coal SOEs). Followed by a series of state-led programmes on nation-wide market promotion, the revenue of the tourism industry increased fivefold, from 563.6 in 2007 to 2559.4 million $ in 2013 (ZSB, 2014). It is also reported that tourism contributed to over 100,000 new jobs since 2008, and Zaozhuang as a typical coal region, surprisingly, became one of the most attractive tourism destinations in China (China Daily, 2012).

More importantly, since 2012, the coal industry began to benefit from the tourism boom. First, the social burden on the coal SOEs was eased, as the emerging industry substantially absorbed laid-off workers. Second, with the improvement of the image of the region, the tourism industry, as well as its sponsors -the local coal SOEs-started to benefit from an influx of inward investment. New branches external to the coal industry were set-up within and by local coal SOEs, leading to a process of business diversification within the coal-mining industry. The general manager of one interviewed coal SOE gave an example: 'because of the tourism, many investors, from real estates, hotels and transportation, made cooperative contracts with us, and now we are not a pure coal firm anymore' (Interview, translated from Chinese). In this sense, Zaozhuang has strategically created a complementary situation between new path development and old path expansion, meaning that adaptation and adaptability go hand-in-hand, positively coexisting and co-evolving by institutional cooperation and benefit sharing.

6.4.2 Fuxin

Fuxin was strongly characterised by a socialist economy, monopolised by one single centrally controlled SOE—Fuxin Coal Mine (FCM)— during Mao's era. In the aftermath of the 1990s' market reforms, its annual GDP growth rate was only 1.1%, significantly lagging behind the national growth rate of 7.2% as a whole (FDRC, 2012). By 2000, the privatisation reforms on FCM caused over 50% employment loss of all residents and hence sparked severe social unrest. Moreover, the verge of coal exhaustion aggravated the already bad situation, which directly raised anxieties among many national state leaders. In this context, re-employing laid-off workers and maintaining regional livelihoods became not only the first priority task for the region, but also the bounden duty for the central state (Hu, 2014). In 2002, the central state, rather than the local, decided to nurture the agriculture sector in Fuxin, as it did have rich arable land resources. That decision, however, was in contrast to the pro-industrialisation mainstream in China at the time and was also more serving employment/social stability. The central state expected tens of thousands of unemployed miners to find jobs in the labour-intensive agricultural sector (China Daily, 2002). To do this, national authorities subsidised leading domestic agricultural firms, such as, Shuanghui and Yili, to invest in Fuxin. The local state was keen on facilitating the top–down decision. Local tax-free allowance, cheap land loans and free farming training programmes were quickly initiated in Fuxin. Within 3 years, a total of 41,500 laidoff FCM miners moved from urban factories to rural farmlands, and found their jobs in agriculture (Interview with Fuxin Development and Reform Commission).

However, the rescue plan proved too difficult to be sustainable. The deregulation of cross-regional M&As in 2005 opened the door for existing path continuity. For instance, FCM acquired a huge open cast mine plant in Inner Mongolia, with an estimated annual coal output of 31 million tons. This, consequently, sparked an unprecedented return of former FCM miners from farming, as they could earn higher incomes and benefit from better social welfare and social status in coal SOEs. The agricultural industry, therefore, faced a dramatic decline. Therefore, this form of adaptability only generated a temporary effect, failing to induce a sustainable process towards structural change.

The old path expansion by cross-region M&As saved the Fuxin economy from decline. But the coal specialisation continued, leading to a severe lack of momentum, and even a strong resistance, to diversify its economy. This ran counter to the central ideal of the 'revitalisation' of the region since 2003. The rising worries in Beijing about Fuxin led to another round of top–down policy interventions. In 2005, soon after the release of the Chinese Renewable Energy Law, Fuxin was designated as a pilot city for wind power industry development. By aligning with China's largest wind power SOE developer Huaneng, the National Energy Bureau started a large-scale (500 WM) project in Fuxin (NEB, 2005). Fuxin viewed this top–down intervention as a chance, and proactively initiated new institutions and wind power favoured policies. A senior official in FDRC pointed at the importance of national policies to Fuxin:

> Our history of centrally controlled economy makes all of us (cadres) realize that every big decision should be made with care, and it's therefore better to listen to the top's idea. We are happy when new national policies come, for developing wind power, we initiated a unique department specializing wind power governance. (Interview, translated from Chinese)

Additionally, with the adoption of national preferential policies such as local content requirements, price regulation and feed-in tariffs, Fuxin quickly became a hot spot for wind power investors in China. An unprecedented firm agglomeration occurred. All China's centrally owned power developers, together with leading wind equipment manufacturers (for example, Goldwind), operated their businesses in Fuxin (Interview). Within a few years, the region significantly restructured the source of electricity production, from nearly 100% reliance on coal before 2005 to about 30% on wind in 2012 (FDRC, 2012).

However, wind power had little impact on the coal industry. The wind power industry is vertically controlled and operated by central SOEs, and its market is barely local. Detached from both the local economic and governance system, the new industry is, to a large extent, running in danger of being a 'cathedral in the desert' (see Table 5.1, Example B1). As a policy-driven and central-state-sponsored industry, wind power in Fuxin is particularly vulnerable to policy changes. It would be 'gone with the [policy] wind', as one municipal official critically stated, and 'would be a waste of local public resources' (Interview, translated from Chinese). Nevertheless, local officials are still supportive of wind power, partly because developing wind power does not collide with interests of the old coal path. Moreover, the local enthusiasm for wind power conveys a political signal, by which Fuxin can enhance the relationship with Beijing, hoping for more future policy benefits. This can be evidenced by the vivid articulations of the Director of Fuxin New Energy Bureau:

> The coal industry is our pillar industry, which contributes to over 70% of GDP growth, while wind power is like a desert. Despite little impact on the local economy, it [wind power] means a lot. It is a clue originally from Beijing, we have to do it and turn it into a success. Only by doing so the central state can trust us [the local state] and we could enjoy more national preferential policies in the future. (Interview, translated from Chinese)

6.5 Comparative Discussion

Based on the accounts presented above, we are able to explore the question of why Zaozhuang is more resilient than Fuxin from the adaptation and adaptability perspective (see Table 6.2). In the case of Zaozhuang, adaptation unfolds, and adaptability emerges, in a relatively endogenous and bottom up way. The local state and the newly-appointed mayor acted as key agents of both adaptation and adaptability. They strategically allowed the extension of the coal industry via cross-region M&As, but at the same time mobilised assets from such path extension to foster the emergence of the tourism industry. More importantly, evidence shows that the coal-SOEs-supported tourism industry played an enabling role in setting in motion the diversification of the coal industry. In other words, Zaozhuang's adaptability is derived from the local endogenous adaptation process, and is oriented to change the existing form of adaptation. Therefore, a reciprocal relationship between adaptation and adaptability can be observed in Zaozhuang (as Type C in Table 6.1).

Such relationship is formed by a strong local agency. The agency may result from the privatisation of ZCM, by which the local state grasped more power by realigning economic assets. But such agency is underpinned by a particular structure of the historically conditioned political economy. The absence of centrally controlled SOEs in Zaozhuang makes the regional economy more independent and autonomous when it comes to responding to crises. By enjoying a loose institutional niche, local actors were able to exercise effective leadership (Hu & Hassink, 2017). This leads to a complementary relationship between adaptation and adaptability, enabling the Zaozhuang economy to renew and transform, and hence to be resilient in the long run. In the case of Fuxin, in contrast, both the selection of new industries to develop, as well as the dynamism of the old coal industry, has been much more at the mercy of top-down policy interventions. These national policy interventions, however, have often been shortsighted, due to a lack of awareness and knowledge of place-specific conditions. Centrally selected new industries (for example, agriculture), for instance, were not created for local structural change, but rather for a short-term remedy to relieve the unemployment burdens caused by the old mining path (refer to Example B2 in Table 6.1). Likewise, the creation of wind power is directly enabled by exogenous agencies (central power SOEs), separated from the adaptation of coal mining. Over time, the separated relationship between adaptation and adaptability started to shift into a competing form (Type A and Type B). This is particularly the case, because the central state continually managed, planned and operated local development, which constrained the local endogenous capacity and creativity, leading to an institutionally hegemonic and industrially incoherent milieu in Fuxin. In such a situation, adaptation and adaptability tend to be disconnected, running in a danger of negative conflicts that might undermine the overall restructuring process.

Why did local actors in Fuxin often act in accordance with policy orders from above? It is because of local awareness of the limited local capacity, in terms of

Table 6.2 Comparison of adaptation and adaptability in Zaozhuang and Fuxin since 2000

	Adaptation	Adaptability	Their relationship
Zaozhuang	Strong path dependence, old path continuity and extension	Endogenously supported; Benefits adaptation (coal-mining); Towards long-run structural change	Mutually complementary; Transformative and robust (Type C)
Fuxin	Short-term path creation (agriculture) (2000–2004) and soon locked in old path continuity (2005–present)	Emergent and temporary (agriculture); Exogenously enabled; disconnected with adaptation; Little impact on structural change	Adaptation and adaptability generally run in parallel (Type B). Adaptability (wind power) and adaptation tend to harm each other (Type A)

decision-making and resource mobility. This awareness stems from several factors at different scales, such as the monopoly of higher-ranking FCM, the well preserved 'top-down' institutions, and even the 'centrally-planned' ideological mindset of local stakeholders. In such a situation, local firm and state actors merely comply with 'top' policies, without sufficient 'place-based' considerations and reflections for long-term development.

Our empirical findings suggest that in the Chinese context, particularly in OICs where SOEs dominate, state agencies seem to be the core in shaping the ways in which adaptation and adaptability are related to each other. And regional resilience, to a large extent, is contingent upon how state and semi-state agents responsively read, strategically interpret and purposively act, which is affected by multi-scalar structural (for example, historically conditioned institutional niches) and contextual (changing, uncertain and emergent contexts) factors (as discussed above). Some regions, such as Fuxin, fail to generate new visions and to take initiatives by their own, not merely because of a lack of endogenous capacities and resources, but more because of limited institutional room for action constrained by broader power relations. Their regional dynamics tend to be mainly at the mercy of top-down policy support and direct interference, leading to a separated and even competing process of industrial evolution. Other regions, such as Zaozhuang, enjoy more institutional flexibility for local experimentations and entrepreneurship, because local actors are aware of their power 'position' in the geographical political economy (Hu & Hassink, 2017; Huang, 1996). They are thus able to purposely manage adaptation towards the emergence of adaptability, based on their own localised conditions and interests.

We thus argue that, in the Chinese context, the way in which adaptation and adaptability interact is more likely to be contingent upon how local officials read and interpret their situated particular positionalities of local-central relations, as well as on how they adapt and govern the local political economy to changing national agendas. Although we recognise that local factors, among others, play a role in affecting a region's resilience, we think that in an authoritarian context macro, structured institutional impact factors matter more for the spatial variations of adaptation–adaptability interactions. These factors, in our view, should be put at the core of uneven regional resilience research in China.

6.6 Conclusions

This article has sought to develop a comprehensive understanding of long-term regional economic resilience. We argue that in the conceptualisation of regional resilience there is a need to avoid dualistic thinking and adopt an ontologically coherent evolutionary perspective. In doing so, we reject the trade-off idea of adaptation and adaptability, and rather address them in a dialectical way.

It is argued that regional economic resilience should not be restricted to short-term recoverability aftershocks or the ability related to resistance. It should be regarded as a two-dimensional long-term notion including resilience processes related to adaptation and resilience capacity related to adaptability. Such adaptation involves a variety of actors at multi-scalar spatial levels, and of heterogeneous actions and interests. It therefore not only seeks to sustain old ways of economic organisation, but also can breed dynamics for structural change. Adaptability is a result, and also an integral part of the adaptation process. In a long-run resilience perspective, it should be an adaptive ability affecting regional economic restructuring.

Furthermore, we have suggested that there are three main inter-related types of adaptation-adaptability relationships, namely *competing, separated* and *reciprocal* relationships. These relationships can be indicative of the geographically uneven processes, patterns and degrees of resilience. To illustrate uneven resilience of regions, we applied this typology to two coal-mining regions in China, namely Zaozhuang and Fuxin. Based on an in-depth comparison, we conclude that Zaozhuang is potentially more resilient than Fuxin. We highlight that the mechanism of adaptability is the outcome of path-dependent adaptation processes related to the power interactions and relations between different actors at multiple scales (Dawley et al., 2015; Essletzbichler, 2012). Our empirical analyses shed light on the important but under-explored role of state agencies in affecting the process and capability of regional resilience in the long run (Bristow & Healy, 2014; Pike et al., 2010).

The findings in this article raise three key issues that need further research. First, conceptually, we suggest that the definition of 'adaptation' needs to be revisited in relation to changes within pre-existing paths and adaptability involving new path creation. In a long-term perspective, new paths may be created as a strategic response to cope with old problems. These new paths might be disconnected to existing paths, less embedded into the local context, and might sustain for a long period of time. Therefore, new path creations should not always be regard as positive to regional structural change (Hu, 2017). Whether being positive or not should be contingent on how they are related to old existing structures of adaptation and, to what extent they are embedded in the whole regional industrial system.

In our view, high-quality regional resilience refers to a complementary, symbiotic, place-based relationship between adaptation and adaptability. Regional resilience is not merely about the ability to overcome shocks. It is rather about long-term endogenous (or endogenously and exogenously well balanced) quality, transformability and sustainability of regional evolution and development (Martin & Sunley, 2015). Resilience does not simply mean a stronger adaptability in terms of creating new industrial paths, but rather, a more 'institutionally coherent', 'inter-industry reciprocal' and 'multi-scalar' adaptation leading to regional structural change.

Second, empirically, we suggest that the roles of state agency, institutional environment and macro impact factors related to evolving geographical political economies are essential for regional resilience research in China. The geographical differentiation of local-central politics, for instance, should be further explored, when analysing long-run resilience in Chinese OICs. Our comparative study confirms the argument that in China's OICs adaptation strategies and practices of firms and industries are not merely affected by specific local socioeconomic contexts, but are largely shaped by their historically embedded central-local politics, as well as by changing national development agendas (Hu & Lin, 2013). We suggest future research on exploring how resilience processes and capabilities are (re)shaped by a combinational force of initial conditions, state agencies, and institutions at multiple scales.

Last but not least, this research generates several policy suggestions regarding the restructuring of OICs. Compared to other types of regions, OICs bear much heavier historical burdens, often following a path-dependent incremental evolutionary process. Policy-makers should avoid a quick policy fix by, for instance, instrumentally adding new industries onto the old industrial systems (Coenen et al., 2014). Future policies should focus more on how to intentionally cultivate endogenous ability of local firms and industries to develop place-specific strategies, rather than to directly transplant new, 'planned' industries/paths. Policies do not necessarily have to focus on developing a completely new economic structure in OICs based on attracting extra-local resources, but may be able to remain competitive through a recombination process of local and extra-local knowledge, assets and competencies (Bathelt et al., 2013; Isaksen & Trippl, 2016). In order to promote healthy and robust resilience in regions, fine-tuning and place-based policies are critically needed. They also need to make sure that new industrial paths are not simply related to, but also can benefit from, and contribute positively to the dynamics of old paths and structures.

Acknowledgements Robert Hassink has made great contributions to this chapter. I express our heartfelt thanks to him.

Note

1. 'Red tourism' is a nation-wide tourism project in China in which Chinese people visit locations with historical significance to Chinese Communism. The central state began actively supporting red tourism in 2005 to promote both the Communist Party's power and economic development, particularly in less-favoured regions.

References

Adger, N. (2010). An interview with Neil Adger: resilience, adaptability, localization and transition. Transition Culture. Available at: http://transitionculuture.org/2010/03/26/an-interview with-neil-adger-resilience-adaptability-localization-and-transition/. Retrieved May 30, 2016.

Arbuthnott, A., Eriksson, J., & Wincent, J. (2010). When a new industry meets traditional and declining ones: An integrative approach towards dialectics and social movement theory in a model of regional industry emergence processes. *Scandinavian Journal of Management, 26*, 290–308.

Bathelt, H., Munro, A. K., & Spigel, B. (2013). Challenges of transformation: Innovation, rebundling and traditional manufacturing in Canada's technology triangle. *Regional Studies, 47*, 1111–1130.

Boschma, R. (2015). Towards an evolutionary perspective on regional resilience. *Regional Studies, 49*, 733–751.

Bristow, G., & Healy, A. (2014). Regional resilience: An agency perspective. *Regional Studies, 48*, 923–935.

Cambridge Journal of Regions, Economy and Society (2015) Local Growth Evolutions: Recession, Resilience and Recovery (Special Issue). *Cambridge Journal of Regions, Economy and Society, 8*, 141–377.

Chapman, K., MacKinnon, D., & Cumbers, A. (2004). Adjustment or renewal in regional clusters? A study of diversification amongst SMEs in the Aberdeen oil complex. *Transactions of the Institute of British Geographers, 29*, 382–396.

Chien, S. S. (2007). Institutional innovations, asymmetric decentralization, and local economic development: A case study of Kunshan, in post-Mao China. *Environment and Planning C, 25*, 269–290.

China Daily. (2002). City hopes miners can become farmers. *China Daily*.

China Daily. (2012). Tourist attractions come at a high price. *China Daily*.

Christopherson, S., Michie, J., & Tyler, P. (2010). Regional resilience: Theoretical and empirical perspectives. *Cambridge Journal of Regions, Economy and Society, 3*, 3–10.

Coenen, L., Moodysson, J., & Martin, H. (2014). Path renewal in old industrial regions: possibilities and limitations for regional innovation policy. *Regional Studies, 49*, 850–865; Cross, R., Grinfield, M., & Lamba, H. (2009). Hysteresis and economics. *Control Systems Magazine 29*, 30–43.

Cross, R., Grinfield, M., Lamba, H. (2009). Hysteresis and economics. *Control Systems Magazine, IEEE*, 29, 30–43.

Dawley, S., MacKinnon, D., Cumbers, A., & Pike, A. (2015). Policy activism and regional path creation: The promotion of offshore wind in North East England and Scotland. *Cambridge Journal of Regions, Economy and Society, 8*, 257–272.

Dawley, S., Pike, A., & Tomaney, J. (2010). Towards the resilient region? *Local Economy, 25*, 650–667.

Economist. (2015). Pollution: The cost of clean air. The Economist.

Essletzbichler, J. (2012). Evolutionary economic geographies. In T. J. Barnes, J. Peck, & E. Sheppard (Eds.), *The Wiley-Blackwell companion to economic geography* (pp. 183–198). Wiley-Blackwell.

Feldman, M. P. (2007). Perspective on entrepreneurship and cluster formation: Biotechnology in the US Capitol region. In K. R. Polenske (Ed.), *The economic geography of innovation* (pp. 241–260). Cambridge University Press.

Foster, K. A. (2007) A Case Study Approach to Understanding Regional Resilience. Working Paper 2007, 08. Institute Urban and Regional Development.

Fuxin Development and Reform Commission (FDRC). (2012). The revitalization of Fuxin in the past decade (in Chinese). *Government Report*, 2012.

Frenken, K., Van Oort, F., & Verburg, T. (2007). Related variety, unrelated variety and regional economic growth. *Regional Studies, 41*, 685–697.

Gong, H., & Hassink, R. (2017). Regional resilience: The critique revisited. In T. Vorley & N. Williams (Eds.), *Creating resilient economies: Entrepreneurship, growth and development in uncertain times* (pp. 206–216). Edward Elgar.

Grabher, G. (1995). The disembedded regional economy: The transformation of East German industrial complexes into Western enclaves. In A. Amin & N. Thrift (Eds.), *Globalization, institutions and regional development* (pp. 177–195). Oxford University Press.

Grabher, G., & Stark, D. (1997). Organizing diversity: Evolutionary theory, network analysis and postsocialism. *Regional Studies, 31*, 533–544.

Hansen, N. (1992). Competition, trust, and reciprocity in the development of innovative regional milieu. *Papers in Regional Science, 71*, 95–105.

Hassink, R. (2010). Regional resilience: A promising concept to explain differences in regional economic adaptability? *Cambridge Journal of Regions, Economy and Society, 3*, 45–58.

Hu, X. (2014). State-led path creation in China's rustbelt: The case of Fuxin. *Regional Studies, Regional Science, 1*, 294–300.

Hu, F. Z., & Lin, G. C. (2013). Placing the transformation of state-owned enterprises in North-east China: The state, region and firm in a transitional economy. *Regional Studies, 47*, 563–579.
Hu, X., & Hassink, R. (2017) Place leadership with Chinese characteristics? A case study of the Zaozhuang coal-mining region in transition. *Regional Studies, 51*, 224–234.
Hu, X. (2017). From coal-mining to coal-chemicals? Unpacking new path creation in an old industrial region of transitional China. *Growth and Change, 48*, 233–245.
Huang, Y. S. (1996). Central-local relations in China during the reform era: The economic and institutional dynamics. *World Development, 24*, 655–672.
Ibert, O., & Schmidt, S. (2014). Once you are in you might need to get out: Adaptation and adaptability in volatile labor markets—the case of musical actors. *Social Sciences, 3*, 1–23.
Isaksen, A. (2015). Industrial development in thin regions: Trapped in path extension? *Journal of Economic Geography, 15*, 585–600.
Isaksen, A., & Trippl, M. (2016). Exogenously led and policy-supported new path development in peripheral regions: Analytical and synthetic routes. *Economic Geography*. https://doi.org/10.1080/00130095.2016.1154443
MacKinnon, D. (2012). Beyond strategic coupling: Reassessing the firm-region nexus in global production networks. *Journal of Economic Geography, 12*, 227–245.
MacKinnon, D., & Derickson, K. D. (2013). From resilience to resourcefulness: Acritique of resilience policy and activism. *Progress in Human Geography, 37*, 253–270.
Martin, R. (2012). Regional economic resilience, hysteresis and recessionary shocks. *Journal of Economic Geography, 12*, 1–32.
Martin, R., & Sunley, P. (2006). Path dependence and regional economic evolution. *Journal of Economic Geography, 6*, 395–437.
Martin, R., & Sunley, P. (2015). On the notion of regional economic resilience: Conceptualization and explanation. *Journal of Economic Geography, 15*, 1–42.
Martin, R., Sunley, P., Gardiner, B., & Tyler, P. (2016). How regions react to recessions: Resilience and the role of economic structure. *Regional Studies, 50*, 561–585.
Morgan, K. (2013). Path dependence and the state. In P. Cooke (Ed.), *Reframing regional development* (pp. 319–340). Routledge.
National Energy Bureau (NEB). (2005). Some suggestions on accelerating the localization of the wind power industry, National Policy Document No. 1953.
Peck, J., & Theodore, N. (2015). *Fast policy: Experimental statecraft at the thresholds of neoliberalism*. University of Minnesota Press.
Pendall, R., Foster, K. A., & Cowell, M. (2010). Resilience and regions: Building understanding of the metaphor. *Cambridge Journal of Regions, Economy and Society, 3*, 71–84.
Pike, A., Dawley, S., & Tomaney, J. (2010). Resilience, adaptation and adaptability. *Cambridge Journal of Regions, Economy and Society, 3*, 59–70.
Regional Studies. (2016). Resilience revisited (Special Issue), *Regional Studies, 50*, 577–656.
Robinson, G. M., & Carson, D. A. (2016). Resilient communities: Transitions, pathways and resourcefulness. *The Geographical Journal, 182*, 114–122.
Sensier, M., Bristow, G., & Healy, A. (2016). Measuring regional economic resilience across Europe: Operationalizing a complex concept. *Spatial Economic Analysis, 11*, 128–151.
Simmie, J., & Martin, R. (2010). The economic resilience of regions: Towards an evolutionary approach. *Cambridge Journal of Regions, Economy and Society, 3*, 27–43.
Smit, B., Burton, I., Klein, R., & Wandel, J. (2000). An anatomy of adaptation to climate change and variability. *Climate Change, 45*, 223–251.
State Council. (2013). Sustainable Development Planning on Resource-based Cities in China (2013–2020) (in Chinese). Available online at. www.gov.cn/zwgk/2013-12/03/content_2540070.htm. Retrieved July 13, 2016.

Tomlinson, P. R., & Branston, J. R. (2014). Turning the tide: Prospects for an industrial renaissance in the North Staffordshire ceramics industrial district. *Cambridge Journal of Regions, Economy and Society, 7*, 487–507.

Wang, M., Cheng, Z., Zhang, P. Y., Tong, L. J., & Ma, Y. J. (2014). *Old industrial cities seeking new road of industrialization*. World Scientific Publishing.

Zaozhuang Statistical Bureau (ZSB). (2014). *Zaozhuang statistical yearbook*. China Statistical Press.

Zhang, J., & Peck, J. (2016). Variegated capitalism, Chinese style: Regional models, multi-scalar constructions. *Regional Studies, 50*, 52–78.

Chapter 7
Institutional Change and Divergent Economic Resilience: Path Development of Two Resource-Depleted Cities in China

7.1 Introduction

The 2008 global financial crisis has negatively hit China's economy. Resource-driven cities have suffered a severe setback (Economist, 2014). Since 2008, 69 resource cities have been selected as 'pilot cities' through a national strategy (State Council, 2013). This strategy aims to help them to cope with resource crises and to enhance resilience. While the resilience of resource depleted cities in China is of great policy importance, it lacks comprehensive and updated scholarly exploration.

Resilience is not only about the ability to absorb and recover from shocks, but is also emphasised as an ability to create new paths (Boschma, 2015). Resilience research in China has tended to draw a taken-for-granted conclusion that those regions are resilient where new industries emerge (Hu & Hassink, 2017a, 2017b). This perception is, however, problematic. Evolutionary geographers have asserted that regional new industries tend to grow out of existing ones by firm branching, in which agents explore technologically related assets in existing industries (Boschma and Frenken, 2011). Exogenous sources, such as national policy intervention, labour dynamics and global production networks, also contribute to regional new path development (Dawley, 2014; MacKinnon, 2017; Trippl et al., 2018; Yang, 2009). In our view, resilience refers to an evolutionary process by which agency, institutions and industries interactively adapt and co-evolve over time. On the one hand, regional resilience links to the economy's structure, performance and overall functioning, reflecting system capabilities of resistance, recovery and robustness (Martin & Sunley, 2015). One the other hand, resilience evolves through agents continuously interacting with a variety of multi-scalar institutions and changing

(This chapter is a reprint version from: Hu, X., & Yang, C. (2019). Institutional change and divergent economic resilience: Path development of two resource-depleted cities in China. Urban Studies, 56(16), 3466–3485.)

X. Hu, *Uneven Economic Resilience of Old Industrial Cities in China*, Urban Sustainability, https://doi.org/10.1007/978-981-99-9279-9_7

contexts (Bristow & Healy, 2014). The ways in which agency and institutions interplay involving dynamic mechanisms of institutional change can generate divergent patterns of path development and exhibit various properties of resilience such as path reorientation and renewal, not necessarily limited to path dependence or creation (Evenhuis 2017a; Isaksen et al., 2018; Martin, 2012).

This article is to reiterate Martin and Sunley's (2015) point that resilience should be conceptualised as evolutionary processes of geographically uneven economic evolution in a long-run context. Resilience differs in space where distinctive agents, institutions and histories are embedded in place-specific socio-economic contexts (Hassink, 2010). It can be compatible with, and understood by, the institutional change approach by particularly focusing on the role of agency—which has been not been thoroughly addressed in the literature (Bristow & Healy, 2014; Lang, 2012; MacKinnon et al., 2009; Sjöstedt, 2014). Moreover, given the conception of path as an on-going dynamics process echoing the nature of economic resilience (Martin, 2010), new research directions have been identified through incorporating notions in evolutionary economic geography (EEG), particularly path creation (Boschma, 2015; Evenhuis, 2017a). Nevertheless, existing literature lacks scrutiny on the causal relationships between institutional change, path development and regional resilience. Moreover, studies on the topic in the recent literature have been mainly conducted in Europe (MacKinnon et al., 2019). Little in-depth research has been conducted at a subnational level in a comparative setting in non-western emerging contexts, such as China (Evenhuis, 2017b).

This article aims to explore the economic resilience of cities by combining the EEG concept of path development with an institutional change perspective highlighting the role of agency. Empirically, taking two Chinese resource-depleted cities as cases, namely, Zaozhuang in Shandong Province and Fuxin in Liaoning, we can illustrate how different modes of institutional change occur in the cities, and how they generate divergent patterns of regional industrial path development that lead to uneven resilience. The remainder of the article is organised as follows. The second section provides a review on regional resilience and path development and identifies research gaps. The third section proposes a conceptual framework to address how modes of institutional change with different types of agents and forms of path development can be logically inter-related to shape resilience. After introducing the study area and research design, the fourth section utilises the proposed conceptual framework to comparatively explore mechanisms of industrial path evolution in Zaozhuang and Fuxin since 2005. The article concludes with theoretical reflections.

7.2 Economic Resilience and Path Development in Cities and Regions: Identifying the Missing Link

Regional resilience refers to a local economy's 'capacity to withstand or recover from market, competitive and environmental shocks' (Martin & Sunley, 2015: p. 3). Different resilience types with various units of analysis and perceptions have been identified (Evenhuis, 2017b). Thus far, most of the research—often equipped with quantitative methods—has focused on how regional economies respond to macroeconomic shocks. Influenced by the idea of engineering and ecological resilience, and by the conventional EEG notions of path dependence and locks-ins (Grabher, 1993), this research strand regards resilience as the short-term actual adaptation in a post-shock context (Yamamoto, 2011). It ties to the firm-based centrality of 'recoverability' and 'resistance' of an economic system, highlighting the equilibrium thinking of a self-reinforcing mechanism for path continuity (Hu & Hassink, 2017b; Simmie & Martin, 2010). In other accounts, resilience refers to the underlying capacity (adaptive capacity or adaptability) of creating multiple possibilities for structural change in the economy (Bristow & Healy, 2014; Evenhuis, 2017b; Martin & Sunley, 2015). In this understanding, resilience entails a non-linear, long-term evolutionary conception, which may go beyond the 'bounce-back' threshold and lead to reorganisation (if they result in alternations in a regional economy, which are purely internally focused) and/or reorientation (when they are directed to the connections of a region with other parts of the world) (Martin, 2012: pp. 11–13). As stated by Evenhuis (2017b: p. 5), evolutionary notion of resilience is developed on the assumption that regional economies are 'never in any type of equilibrium', but instead are in 'a dynamic process of constant renewal'. In evolutionary resilience, the focus is 'on the long-term evolution of regions and their ability to adapt and reconfigure their industrial, technological and institutional structures in an economic system that is restless and evolving' (Boschma, 2015: p. 735).

Despite valuable evolutionary conceptualisations, the existing regional resilience research has tended to avoid path dependence, to overlook micro-level dynamics and logics and even to simplify resilience to an ability for industry path creation (Boschma, 2015; Henning et al., 2013). The trade-off idea between adaptation and adaptability in resilience, namely, adaptation as path dependent changes with the existing path undermining adaptability as path-breaking structural transformation and vice versa, exemplifies the point (Hu & Hassink, 2015, 2017b; Pike et al., 2010). The adaptability-favoured accounts indicate that broader evolutionary concepts, such as alternative path dependence, co-evolution and complex adaptive systems (CASs), emphasising micro-foundations of (often) incremental dynamics and mechanisms, have been insufficiently incorporated into regional resilience research (Boschma, 2015; Bristow & Healy, 2014; Evenhuis, 2017b; Gong and Hassink, 2019).

Recently, EEG scholars have started to highlight the role of agency and institutional change in conceptualising regional resilience. Bristow and Healy (2014) take a CAS approach to regional resilience by acknowledging the variable capabilities of agents, and propose three key points: (1) agents not only anticipate but also adapt or transform their activities to certain contexts; (2) the agency of agents (e.g., capability and purpose) is shaped by their knowledge of the environment, as well as by broader agendas; and (3) individual agency can gradually form collective agency and impact system-level institutions. More recently, a growing body of literature in EEG, albeit not directly linking resilience, has elaborated on agency as a motive for institutional and economic evolution through the notion of regional development paths. Sotarauta and Suvinen (2018) argue that micro-level institutional agency is the key to understanding the upward causation of macro-level path creation. Such institutional agency, if in a collective form, can be tied to the notions of place leadership (Hu & Hassink, 2017c) or innovative entrepreneurship (Grillitsch & Sotarauta, 2019), which are able to effectively work across different organisational boundaries, exploit institutional ambiguities and strategically facilitate social change, cumulatively enabling system renewal (Steen & Hansen, 2018). Similarly, Isaksen et al. (2019), in discussing the micro-level mechanisms of economic restructuring, distinguish between path extension and upgrading (based on individual agency impacting firm institutions) and path diversification and creation (enabled by system agency shaping structural institutions). These studies allow us to argue that resilience is about institutional change processes—mediated by human agents—that shape regional industrial paths for economic renewal (Evenhuis, 2017b; Gherhes et al., 2018; MacKinnon et al., 2019).

We appreciate the valuable agency-based resilience research focusing on institutional dynamics in relation to regional path development. However, it is still unclear how the notions of agency, institutional change, path development and resilience are causally related to each other. In Fig. 7.1 we illustrate the basic causal chain of these notions. First, individual (firm, state and other organisational) agents intentionally adapt to multi-scalar contexts. They show bounded capabilities and heterogenous purposes, which shape their embedded institutions mainly at the firm/ industry/ cluster level. Their agencies (at least in the short term) thus have limited impact on system institutions. Individual agents may work beyond organisational boundaries and seek broader cooperation and consensus. They may form a collective agency aiming at higher system-level institutional change. Second, firm- and system-level institutional change can both impact human behaviour in the regional economic context. While firm-level institutional change may merely generate certain sector-wise impacts, system-level change can significantly influence the overall economic structure in which main industrial development paths begin to take shape. Third, development paths represent main economic activities with regionally distinctive characteristics. They condition the degree of success in economic renewal as a key indicator of regional resilience (Bristow & Healy, 2014). Another research pitfall in the literature is that institutions are poorly defined, and their dynamics are often loosely regarded as either radical or incremental, overlooking the complex sub-type nature of the latter (Streeck & Thelen, 2005). This article calls for a more nuanced understanding of which type of institutional change generates which pattern of path development and resilience in regions. A conceptual framework will be constructed of comparative regional resilience, which this article will introduce below.

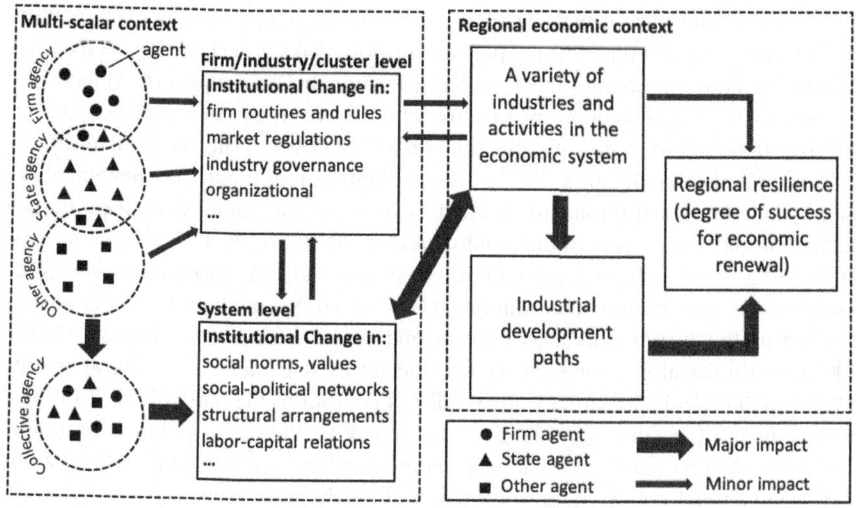

Fig. 7.1. The causal chain of agency, institutional change, path development and regional resilience

7.3 Path Development in the Process of Evolutionary Resilience: An Institutional Change Approach

In a city economy, development paths are conceptualised as an evolving assembly of key industrial evolution processes, entailing varieties of major mechanisms of interplay between change (actions for structural transformation) and continuity (actions for structural stabilisation) (Isaksen, 2015; Steen & Hansen, 2018). The EEG concept of path development can be understood through the institutional change perspective (Bathelt & Glückler, 2014; Martin & Sunley, 2015). For one thing, institutional change is place dependent: agents are embedded in place-specific institutional environment and their behaviors are conditioned by history (Mahoney & Thelen, 2010). Besides this, institutions can be significantly changed by 'institutional entrepreneurs' collectively striving for long-term regional transformation and resilience (Evenhuis, 2017a; Gherhes et al., 2018).

The nature of evolutionary resilience also echoes the logic of institutional change as *incremental* processes in socio-economic systems (Anderies et al., 2004; Berman et al., 2012). As Martin and Sunley (2015: 10) put it: 'Local economic structures do not change completely overnight, even in response to major shocks. Resilience is not an either/or feature or outcome, but a complex process that admits of many possible combinations of change and continuity'. Here, 'many possible combinations of change and continuity' reflects the variegated modes of institutional change in path development. Scholars have developed a typology of institutional change, including layering, conversion, drift and displacement (Boas, 2007; Streeck & Thelen, 2005; Thelen, 2003). Each type means a specific way in which continuity and change interact. Specifically, each refers to the aggregate activities of agents that focus on one major mechanism of path development in an economy, such as path creation, path extension, path branching and path exhaustion (Isaksen et al., 2018). And each is enabled by a type of 'change agent' with one featured behaviour pattern (Mahoney & Thelen, 2010). In a city economy, various types of paths may co-exist and co-evolve, but over time certain types of paths may take lead position (this is particularly the case in organisationally thin and less diversified regions) that can fundamentally shape resilience. Institutions are defined as 'rules of the game' that shape human behaviour and verse visa (Bathelt and Glückler, 2014; North, 1990). They may be formal (e.g., codified laws, regulations, contracts and policies) or informal (e.g., social norms, conventions and routines) (Gertler, 2018). Institutions can be distinguished as the *institutional environment* and *institutional arrangements* (Martin, 2000). Institutional environment means 'the set of structures at a more aggregate level, which provide the larger framework that conditions the particular institutional arrangements' (Evenhuis, 2017a: p. 511). It may be the larger political economic structures and regulations (e.g., labour-capital relations, local-central state relations and supranational policy frameworks), as well as the wider sets of norms, conventions and cultures among people and in places. Institutional arrangements refer to (in)formal organisational regimes and coalitions often enabled by agency, consisting of market institutions primarily at the level of firms/industries and governance institutions primarily at the system (or regional) level (Evenhuis, 2017a; Geels, 2004). They thus may range from industry procedures, agreements and routines to state-led policy initiatives, regulations and strategies. Additionally, institutions may also include broader sets of agents as key institutional components and builders (Gertler, 2018).

We identify two key interrelated factors affecting modes of institutional change, namely, the roles of agency and institutional environment (Zukauskaite et al., 2017). For one thing, scholars stress the important role of existing socio-economic structures in regions (e.g., types of regional innovation systems) in conditioning the regionally divergent institutional environment (the degree of institutional ambiguity and diversity) that shapes (but does not determine) the purpose, power and discretion of agency for resilience (Bristow & Healy, 2014; Evenhuis, 2017b; see also Fig. 2). For another, agents whose purpose, power and discretion towards change are different can be categorised. Different types of agents interpret, interact with and shape the institutional environment in different ways and thus may generate different modes of institutional change. The latter underpin the emergence of featured development paths (Table 7.1) in uneven regional resilience (Fig. 7.2).

(1) *Layering*

Layering refers to an adding process of new institutions on top of or alongside existing ones (Streeck & Thelen, 2005). Change agents may come in two varieties: *mutualistic symbionts* and *subversives* (Mahoney & Thelen, 2010). Mutualistic symbionts are embedded in, benefit from and abide by the status quo in a system. They add new elements for preserving (or contributing to) the robustness of the existing institution (Van der Heijden, 2010). In this mode, agents primarily work for the structural stabilisation of a local economy (Martin, 2012). They may set new firm/industry-based

Table 7.1 Characteristics of institutional change and path development

Change modes	Characteristics of institutional change	Corresponding mechanisms of path development
Thickening	The adding process of new institutions for preserve and/or enhance existing ones	Persistence and extension of an existing industrial path based on the increase of momentum onto existing industries along established institution
Positive layering	The introduction of new rules for reforming the existing institution	Creating new industrial paths based on existing assets, or transplanting new industries from elsewhere
Conversion	The changed enactment of existing rules due to their strategic deployment	Renewal or modernization within an industrial path, or new industries 'branch' out of existing technologically related ones
Drift	The changed impact of existing institutions because of shifts in the environment	An industrial path lacks adaptation and gradually loses momentum and are locked in decline
Displacement	The radical removal of existing rules and the introduction of entirely new oncs	Deconstruction of existing industrial paths and creation of entirely new ones in a radical way

Source Adapted based on Mahoney and Thelen (2010) and Isaksen et al. (2018)

arrangements for investing more capital, labour or infrastructures, expanding scales of production or giving incentives to the old system. This layering, named *thickening*, may lead to path persistence (protecting and sustaining an existing main economic activity and structure) and/or path extension (enhancing an existing industrial path by incremental innovations within existing predominant social and technological system) (Isaksen, 2015; Tödtling & Trippl, 2013). It is likely to occur where regions are organisationally thin and industrially specialised with limited agency, and/or where the institutional environment is rigidly structured by a higher-level power with little ambiguity (Van der Heijden, 2011; Zukauskaite et al., 2017).

Subversive-led layering, however, brings new institutions for shaping the ways in\ which the original institution functions (Thelen, 2003). Agents can, for instance, foster new industries by adding new industry-based arrangements or technologies into existing market institutions. The rise of onsite water recycling technologies in Beijing, China exemplifies how prevailing institutions are shifted by which agents tentatively apply new technologies in a niche market (Binz et al., 2016). This can be regarded as *positive layering*, which enables path creation and/or transplantation, indicating a higher degree of success in economic renewal than thickening (Fig. 7.2). Positive layering may occur in regions where the established institutions are ambiguously or loosely defined. Besides, a certain degree of discretion/power of agents in institutional interpretation and enforcement is needed.

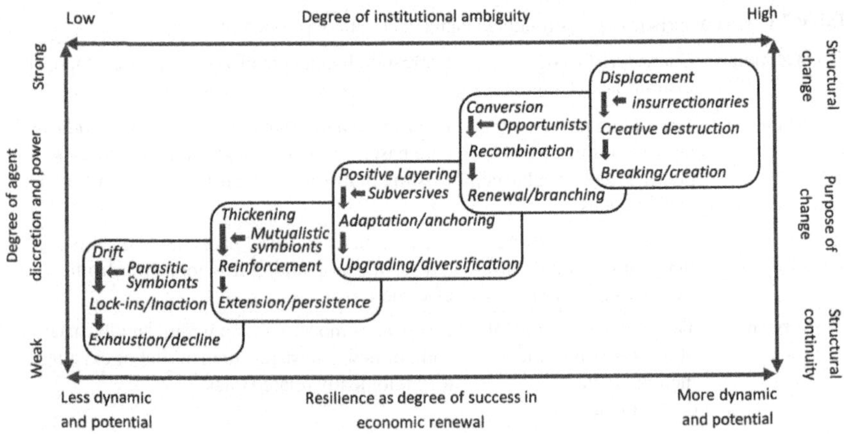

Fig. 7.2 An institutional change framework for understanding divergent resilience of cities and regions

(2) *Conversion*

Conversion involves the redeployment process of old institutions for new purposes (Streeck & Thelen, 2005). Change agents are *opportunists* who practice institutional ambiguities resourcefully (Mahoney & Thelen, 2010). They may 'preserve institutional continuity and exploit whatever possibilities exist within the prevailing system to achieve their ends' (Mahoney & Thelen, 2010: p. 26). Strambach and Halkier (2013) argue that in an overarching institutional setting, there remains leeway for agents to exploit creativity. New industries can be fostered without the need for breaking the system institution, but rather through cumulatively shifting the existing institutional environment through institutional change at the firm/industry/cluster level.

Conversion involves a time-needed process of social interaction. Change agents create new institutional arrangements for dialogue with existing conservative agents. This dialogue involves agency on knowledge exchange, learning, resource complementation and consensus building, such as by way of subjective and symbolic narratives on resilience (Goldstein et al., 2013). For example, Arbuthnott et al. (2010) stress the importance of social interaction between forestry and biorefinery actors in Örnsköldsvik, Sweden. They stress that such interaction has spurred new inter industry contracts and shared strategies towards a recycling economy. In this sense, conversion is associated with the notion of 'relatedness', meaning that industrial agents with social entrepreneurship can facilitate structural change at the system level for regional diversification and transformative resilience (Boschma, 2015; Neffke et al., 2018). This conceptually echoes Grillitsch's (2015) argument that dynamic interaction in different institutional layers of industries can potentially enable knowledge creation and structural dynamics, showing the 'reorganisation' dimension of resilience (Martin, 2012). It means a reuse, recombination and anchoring process of existing institutions, seeking to endogenously reform the existing system. Therefore, conversion enables *path renewal* and *path branching*, showing a more dynamic resilience than layering (Fig. 7.2). Regions with a high degree of organisational relatedness and thickness, and with a certain degree of institutional ambiguities, have advantages to enable conversion (Isaksen et al., 2018). But conversion can also occur in organisationally thin regions if they have capably strong agents and a long-term purpose for structural change.

(3) *Drift*

Drift refers to a situation of changed impact on existing institutions because of shifts in the institutional environment and a lack of response of agents to these shifts (Streeck & Thelen, 2005). It may result from the failure of agents to update an institution. These agents are like *parasitic symbionts* who rely on existing institutions for private gain, lacking motives to impact the institution in any way. Old institutions are not protected, and evolve randomly with external environment dynamics. Drift is a process characterised by high vulnerability, exhibiting a low degree of success in economic renewal. This mode may result in *path exhaustion*, namely, when an economy gradually loses dynamism and eventually faces decay (Trippl & Isaksen, 2016). Self-contained and organisationally thin periphery regions are particularly subject to drift (Isaksen, 2015). Besides, drift may also occur when agents lack the opportunity, power or motivation to change, restricted by multi-scalar networks and rigid structures (Fig. 7.2).

(4) *Displacement*

Through displacement, existing institutions are supplanted by new ones (Van der Heijden, 2010). Change agents are *insurrectionaries*, who reject and do not abide by the existing institution, but strive to displace it with an entirely new one (Mahoney & Thelen, 2010). This may happen when hit by profound shocks, through which the existing system is permanently changed, unable to return to its previous state (Simmie & Martin, 2010). It may co-evolve with and result from Schumpeter-style technological breakthrough derived from the firm/industry level (Gong & Hassink, 2019). It can be also triggered by powerful agents in pursuit of a fundamental reform in the first place, incessantly destroying the old system and creating an entirely new one. This is a type of institutional change leading to path breaking and (radical) path creation. It thus indicates greater power of resilience than any other cases (Fig. 7.2). Despite a very low possibility of occurring, regions whose institutions are highly dynamic and open, and where agents have strong system-level power, discretion and purpose aiming at structural change, are of potential to enable displacement (Fig. 7.2).

7.4 Study Areas and Research Design

Two Chinese resource-depleted cities, namely Zaozhuang and Fuxin, were selected for case studies. First, they have been regarded as less favoured, due to long-term mining specialisations and resource depletion (State Council, 2013). Second, Zaozhuang is located in Shandong Province whose economy is more marketised and decentralised. Fuxin is a peripheral city of Liaoning in Northeast China. Compared to Shandong, Liaoning's economy has been more resource-driven, dominated by central state-owned enterprises (SOEs), and fed by national investment projects over decades. This has led to a regionally unique outcome called the 'Dongbei Phenomenon' (东北现象). This phenomenon reflects a set of well-preserved institutions all heavily reliant upon and accountable to upper states/SOEs, including danwei cultures, local bureaucracy and governance traditions (Hu & Hassink, 2017a). Third, both have recently exhibited distinct development trajectories (Table 7.2). Zaozhuang's annual GDP, with its fast growing tourism sector, grew 4% faster than Fuxin's during the past decade (Hu & Hassink, 2017c). Fuxin has undergone a sluggish development, with 12.2% negative GDP growth in 2016, despite the emergence of new industries such as wind power (FSB, 2017).

The cases are worthwhile to compare, not just because of the distinctive regional institutions, but also because of having different performances with the same nature of crisis (Table 7.2). Qualitative methods are applied based on multiple on-site fieldworks with 60 in-depth interviews between 2013 and 2016. The interviews involve representatives from firms, authorities and organisations (Table 7.2). Secondary data on industrial development from 2005 onwards was collected from statistical yearbooks, newspapers, firm reports and state documents. With the data, we explore how institutional change unfolds, and how these changes impact the industrial path development and economic resilience of the cities. Answering the plea of Evenhuis (2017b) for comparative case studies on resilience in different contexts, this article provides vivid evidence that comparative case study design is the most appropriate for studying the institutional mechanisms and regional path development forms of evolutionary resilience.

Table 7.2 References on case areas and interview information

City	Population/ GDP (billion $) of 2016	Average GDP growth rate (2005–2016)	Miner number change (2005–2016)	Information on interviewees (number of interviewed organizations)
Zaozhuang	3.91 (mil.)/ 34.1	12.28%	−83,400 (decrease)	Mining firm (12); tourism firm (3); state body (6); industry association (2); vocational college (1)
Fuxin	1.77 (mil.)/ 6.45	6.41%	+41,000 (increase)	Mining firms (2); wind power developers (8); wind manufacturers (5); state departments (7)

Source City Statistic Yearbook of Zaozhuang and Fuxin in (2017)

7.5 Understanding the Divergent Resilience of Two Chinese Resource-Depleted Cities

7.5.1 *Zaozhuang: Positive Layering and Conversion Towards Path Creation and Renewal*

In the 1990s, Zaozhuang's economy was dominated by Zaozhuang Coal Mine (ZCM), a provincial SOE. The 1998 SOE reform ended ZCM's domination and split it into several small (sub)municipal-level SOEs. From 2002, coal resources started to dwindle. Anxieties about economic decline spread, but the city still lacked the vision to cope with the crisis. In 2005, a new mayor took office who was determined to transform the economy.

The new leader unusually advocated nurturing tourism. Zaozhuang was the location of a key battle won by China against Japan during the Second World War, and there were also plenty of cultural assets related to the Grant Canal which remained underexploited. Moreover, developing tourism could lighten the social burdens of mining SOEs by absorbing lay-offs (interview with the office director of Zaozhuang City Government, May 2013). Despite these good reasons, the idea encountered wide skepticism from local cadres. This drove the mayor to take an unconventional initiative. He rejected selling lands for transient rewards and rather convinced wider agents to develop tourism. As the director of Zaozhuang Development and Reform Commission (ZDRC) (interview, April 2013) stated: 'He broke unilaterally a 60 million RMB land-selling agreement with a real estate company. He stressed that "selling lands once, destroying the historical values forever"'.

It took two years to turn the initiative into a shared goal. For example, the mayor proposed a vivid concept in his formal political speeches and daily talks in public, namely the 'tank of fuel' theory (an analogy between the resource-based economy and a car running on only a tank of fuel, referring to the critical condition of Zaozhuang for structural transformation). At every chance, he emphasised the need to transform the coal-based economic structure into a more sustainable one. As a result, it helped local stakeholders to accept the initiative and see the possibility of doing things differently. The CEO of a local coal SOE stressed the importance of the theory:

> We [coal companies] never think of new developing sectors unrelated to mining, not mention to tourism. We did need a vision. The 'tank' theory was explained and widely disseminated in many occasions for two years, and eventually, a consensus for tourism development was achieved. (Interview, January 2014)

Based on this, local state elites were able to mobilise local governance resources for tourism development. This can be exemplified by the launch of the Zaozhuang Tourism and Service Industry Commission (ZTSIC) in 2010, and also that of tourism training programmes in two local vocational colleges (ZEC, 2014). These organisation-adding arrangements helped to generate a new labour poor supporting the tourism sector. However, with local coal exhausting and plant closure, the city encountered financial obstacles to fostering tourism development. To overcome it, the mayor set up a new governance rule in the mining industry, namely, *conditionally* approving coal SOEs to enact mergers and acquisitions (M&As) in China's coal-rich regions. The condition was that these SOEs contribute a share of the profits reaped from M&As to support tourism, e.g., by transferring money to the ZTSIC or by financing and/or operating hotels, restaurants or scenic spots (interview with the office chief of ZTSIC, April 2015). As such, the local state acquired initial investment funds to build a tourism theme park—Taierzhuang Ancient Town. For the SOEs, operating coal plants out of the city, on the one hand, meant they survived the local depletion crisis. On the other hand, they became key investors, shareholders and operators of local tourism-related businesses. As the vice manager of a coal SOE (interview, August 2013) recognised: 'The strategy has significantly changed the past monotonous business model of coal companies. It enabled the coal industry to step into other entirely new businesses. Like our company, we are no longer a pure coal company any more'.

In order to explore market sources, the mayor introduced a set of new criteria onto the prevailing GDP-centric local cadre evaluation system. One newly added criterion was the extent to which cadres attract tourists to invest in Zaozhuang. To put it into action, he initiated the 'Zaozhuang Two Day Tour' programme. By giving each (municipal- and lower-level) government unit a compulsory quota of tourists to attract, local stakeholders were designated (and incentivised) to explore new tourism markets. As the Director of Zaozhuang Coal Industry Bureau (interview, May 2013) put it: 'Our bureau got a quota of 3000 tourists in 2012, but we attracted over 5000 tourists from Xuzhou, Baoding, and Hefei combined. Given our performance, many collogues got promotion'.

As a result, between 2008 and 2016, the tourism industry revenue increased sixfold, from RMB563.6 million to 3760.5 million yuan (ZSB, 2017). It has contributed an average increase of 2% annually to the city's GDP growth and significantly shifted the share of the tertiary sector in the economy from 28.4% in 2008 to 41.4% in 2016 (ZSB, 2017). Over 100,000 new jobs were created in Zaozhuang, 80% of which benefited from the tourism and service industry. Moreover, it is estimated that by 2020 tourism will be the city's pillar industry, accounting for over 50% of GDP growth (China Economic Net, 2014). More relevant to our purpose here, not only does the tourism industry itself grow rapidly, but it also starts to benefit the mining industry. It is documented that between 2005 and 2016 about 56,000 of a total 83,400 laid-off mining workers were re-employed in tourism, the others being primarily designated by SOEs to work in newly built mining plants elsewhere (e.g., Shannxi, Yunnan) (ZSB, 2017). Such efforts helped the SOEs alleviate unemployment and paved the way for economic renewal. Zaozhuang is evident as having undergone quality-orientated institutional change. This change was achieved by positive layering and conversion in which local state elites played key roles (Table 7.3). In terms of positive layering, the introduction of new governance institutions including new developmental visions (e.g., developing tourism), concepts (e.g., the 'tank of fuel' theory), policy initiatives and organisations created social dialogues between change (the mayor) and continuity agents (local coal stakeholders). Such dialogues involving incremental social reconstructs have shifted the institutional environment at the broader system context, namely, allowing local agents to be cognitively reflective, to see the context differently, and thus generating collective agency towards new path creation. This new path is underpinned by an endogenous process in which local state elites with strong institutional leadership strategically enable major changes both in governance institutions and institutional environment at the regional level. More importantly, the new industry is not just fostered alongside the old coal industry, but rather has made significant impacts on the latter, evident by considerable inter-industry labour mobility and reciprocity. The revisions on the conventional GDP-centric cadre evaluation system set in motion for coal SOEs to step into tourism, leading to an intra-organisational diversification within the old path. This is a typical 'conversion', in which the established path has been gradually orientated to benefit tourism as a new path, showing a sign of path renewal.

Change agents in Zaozhuang acted both as subversives and opportunists (Table 7.1). They interpreted the existing coal-based institution as a source of development and chose to gradually transform it. To do that, they strategically made good use of the national policy (M&As) as an opportunity to draw resources locally for the new path creation (tourism) and diversification (service sectors). We acknowledge that the strong local agency in Zaozhuang is not inherent, and is rather underpinned by the interests of agents for political promotion and the loose institutional environment (related to the relatively diversified economic structure and the lack of higher socio-political influences) (Hu and Hassink, 2017c). Zaozhuang therefore has entered into a smooth, inter-organisationally interactive and agency coherent way of economic renewal.

Table 7.3 Differences in institutional change, path development and resilience between Zaozhuang and Fuxin

Item city	Mode of institutional change	Type of change agents	Form of path development	Differences in economic resilience
Zaozhuang	Positive layering and conversion (*Serve to new goals and transformative*)	Subversives and opportunists (*Local state leaders*)	Path creation and Renewal (Towards structural change)	More dynamic, (High degree of success in economic renewal)
Fuxin	Thickening (*Obey national policies and stabilizing*)	Mutualistic Symbionts (*National actors and local state elites*)	Path persistence and extension (Towards structural continuity)	Less dynamic, (Low degree of success in economic renewal)

7.5.2 *Fuxin: Institutional Thickening Trapped in Path Persistence and Extension*

Fuxin was one of China's key industrial powerhouses in the pre-reform period. Its prosperity between the 1950 and 1980s represented the epitome of China's socialist industrialisation under top-down state-led investments. However, in the 1990s Fuxin underwent a severe stagnation (Hu & Yang, 2018). Hit by the 1998 radical SOE reform and suffering from coal depletion since 2000, Fuxin's monopolist—Fuxin Coal Group (FCG)—was forced to shut down almost all its plants. Over 200,000 workers were made unemployed, pushing the city to the brink of economic collapse and social disorder (FSB, 2014). Thanks for the national aid measures since 2001 (e.g., Northeast China Revitalisation), Fuxin's economy has recovered.

One key national measure is the M&A strategy as also mentioned in the Zaozhuang case. Despite not being compulsory in implementation, the strategy was seized by the Fuxin government as a 'now-or-never chance' (interview with a municipal official, April 2014). It was widely believed among local shareholders (e.g., state and SOE elites) that the strategy signaled that Beijing could help Fuxin to regain dynamism back onto the coal-driven track of growth. The Vice Manager of FCG (interview, January 2014) stressed that: 'The Fuxin economy has been centrally controlled for decades. Once facing problems that we cannot solve, Beijing will always help us. We therefore all have a strong sense of political proximity to the central state'.

Such a strong upward dependent mentality among Fuxin's state elites has resulted in a local convention of activism subject to central decisions. In 2006, a special local state organisation was established to implement the M&As strategy for FCG. It led to the acquisition of a large open cast mine in Xilingol, Inner Mongolia, and the construction of a special railway directly connecting Fuxin. As such, the coal industry survived and continued to dominate the local economy, regardless of local coal depletion.

Since 2005, Fuxin's economy has been marked by the rapid growth of wind power, despite the dominance of the mining industry (It still accounted for 31.1% of Fuxin's GDP in 2016; FSB, 2017). The driving forces behind the new sector was not local, but instead came from Beijing. As the Office Director of Fuxin Government (interview, February 2015) stressed: 'When Beijing decided to foster renewable industries in China in 2005, Fuxin was designated as a pilot city. This is because Beijing's worries on Fuxin's problematic economy turned out a deliberate policy care'.

In a short time span, the National Energy Administration designated nine centrally owned wind power SOEs to deploy projects in Fuxin. Again, the local state was enthusiastic about the strategy, and opted to fully comply with it. For instance, a new municipal-level bureau, namely the Fuxin New Energy Bureau (FNEB), was launched in 2010. FNEB's duty is to deliver national instruments and to provide tailor-made services (e.g., offering land quotas and subsidies) to the central SOEs. Likewise, the local financial system was regulated to provide low-interest bank loans and tax exemptions specifically to wind power firms. With these changes to governance institutions, Fuxin quickly became a leading wind power generator in China, with its installed capacity of 2094 MW in 2011, and accounted for over 50% of the total capacity of Liaoning (FSB, 2012).

Nevertheless, Fuxin still lacked the dynamism to restructure its economy. Critical voices from our interviews indicated that the so-called 'institutional innovation' of the launch of FNEB (since it is rare among municipal-level bureaus in China) was for the purpose of pleasing central SOEs within the wind power industry, rather than for higher level structural change. As an official in the Fuxin Development and Reform Commission (FDRC) (interview, October 2016) complained: 'We chase central policies and spare no effect to implement them. New institutions are set to respond national polices by no means to change the existing economic structure'.

Moreover, wind power in Fuxin is operated by central SOEs and its consumption market is barely local-based. Most firms focus on equipment assembly lacking original innovation. The lack of local skilled labour and technologies has hindered knowledge spillovers between central SOEs and local firms. Despite huge investment for years, the new industry had only created 5500 jobs by 2015 (China Energy News, 2017). The recent top-down regulation tightening on forestry and land protection has significantly slowed down the approval process of large wind power projects in Fuxin, resulting in a significant decrease of central investment (FBS, 2017). The new industry indeed, has contributed little to the local economy. Similar evidence can be found in other sectors, such as agriculture, food and coal-chemicals, which are all initiated and supported by national policy interventions (Hu & Yang, 2018). However, with the M&A strategy the mining industry grew, as its share of the local GDP increased from 14.8% in 2010 to 23.1% in 2016 (FBS, 2017: Table 12). But the old path seems to have failed to boost the economy too. Due to the shrinking coal markets and prices, Fuxin suffered from a severe economic decline with 5.5% negative GDP annual growth rate in 2015 and an even worse record of –12.2% in 2016 (FSB, 2017). The Director of FDRC (interview, October 2016) noted: 'We [local state elites] focus too much on how to seize national policy opportunities to either pragmatically create new sectors or sustain the preferred mining industry, and thus overlook our own advantages, resources and capabilities'.

Based on the empirical evidence, we found that the resource crisis in Fuxin did not trigger significant changes in the established institution of the economy. Fuxin's change agents include national SOEs and local state elites (Table 7.3). As *mutualistic symbionts*, they had no interest in reforming the institutional status quo of the system. In particular, local agents treat national policies as a code of conduct, and thus act in unison with them, bypassing local-specific needs and conditions. Their actions are not only limited in scope and discretion because of the limited institutional ambiguities (mainly caused by the highly specialised economic structure and strict central-local state relations), but are also constrained by historically conditioned local cultures avoiding risk-taking and entrepreneurship. As a result, change agents choose to narrowly set new governance arrangements for either maintaining their old industry path (by implementing M&As) or blindly catering to top-down decisions without reflective concerns on whether and to what extent they match place-specific conditions. To a large extent, the new industry itself can be seen as an added institution (an industrial layer) that stays in isolation with the established main path—mining. This refers to thickening—the institutions are thickened, whereas the core functionality of the economy remains barely changed.

Besides the economic structure, the long-term presence of central SOEs and national policy intervention has led to a self-reinforcing institutional environment. This environment undermines the power, discretion and purpose of agents towards structural change. Local stakeholders deliberately choose to neglect local comparative advantages (e.g., with only a 120 km distance from both the nearest harbour and Liaoning's capital city, Fuxin could have improved its transportation to these places instead of building the railway to Xilingol), and underrate locally rooted non-coal sectors such as agate polishing/design and hydraulic machinery. This can be seen as a typical political lock-in, which is preventing substantial changes at the system level (Grabher, 1993). The rise of wind power does not demonstrate a robust new path creation. With institutional stubbornness and strong protectionism towards the coal industry, Fuxin's economy is trapped in path persistence and extension, showing a low degree of success in economic renewal (Table 7.3).

7.6 Conclusions

In this article, we have sought to integrate an institutional change approach into the studies on the economic resilience of cities. Economic resilience is an evolutionary notion that deals with micro-level processes and mechanisms (Martin & Sunley, 2015). It is geographically and historically informed by, characterised by and made up from intertwined processes of change and continuity in which multi-scalar agencies and institutions interactively play a role. By linking EEG concepts of path development, this article argues that the resilience of cities can be differentiated through an institutional change perspective highlighting the role of agency. Different modes of institutional change are useful to explain regionally divergent industrial path development that matters for uneven economic resilience.

Our empirical analyses elucidate that in the context of Chinese resource-depleted cities—as one important type of old industrial cities—the state and SOEs at multiple levels play salient roles in affecting governance institutions (rather than market ones) and the institutional environment in relation to regional path development. In cities such as Zaozhuang, change agents are able to shape broader system-level institutions partly because of the entrepreneurial agency with a certain agenda for transformation, and partly because of the lack of national 'hands' (policy interventions and controls). These cities thus can facilitate positive layering and conversion, which underpin path creation and renewal towards structural change. Others like Fuxin may fail to foster institutional change at the system level. Their main activities tend to be at the mercy of and limited by top-down policy interference in a self-reinforcing institutional environment. These cities are likely to experience a thickening of firm- and organisational-level institutions, resulting in path extension and/or persistence with a low degree of success in economic renewal. We contend that an institutional change perspective allows us to scrutinise the variegated mechanisms of path development that comprehensively reflect capabilities, scopes and agendas of resilience in regions and cities. The ways in which institutions and industrial paths co-evolve are not only conditioned by place-specific assets and histories, but are also contingent upon how agents read, interpret and impact multi-level institutions and tempo-spatial dependent changing contexts, highlighting the important role of agency in understanding uneven processes of economic resilience (Martin & Sunley, 2015; Zukauskaite et al., 2017).

Moreover, the study provides additional and comparative insights from China to understand a new development path in old industrial cities (OICs). The case of Zaozhuang illuminates a distinctive mechanism of new path development that is related neither to endogenous technological/industrial branching nor to exogenous industrial importation (Isaksen et al., 2018). Rather, agents gradually forge a social space for communication within the established path/institution, inviting heterogeneous agents to search an updated (also collectively shared) institutional arrangement at the system level for structural change (Bathelt & Glückler, 2014). This can, as shown, lead to intra-organisational diversification and path renewal. Meanwhile, the case of Fuxin, as a typical thin peripheral region, provides evidence that resilience is highly contingent upon exogenous sources and central policy agency (Trippl et al., 2018). Such exogenously-led new path creation, however, may undermine 'reorientation' of the existing institution, if agents forge new paths merely as a transient response to cater for established paths, and if there lacks the agency of a knowledge/resource anchoring process orientated to macro-scalar institutional shifts (Binz et al., 2016). For strengthening the resilience of peripheral and organisationally thin regions, we propose, the priority task is not to develop new industries. Rather, the focus should be on how to incrementally de-lock the core institution from multi-scaled structures, and to explore more institutional ambiguities (or more frequent occurrences of 'structural holes') and diversities for renewal.

To conclude, the article provides a qualitative basis to understand the economic resilience of cities. The institutional change approach with a focus on the role of agency can be a promising way to conceptualise path development and analyse regionally uneven resilience (MacKinnon & Derickson, 2013). However, the approach needs to be further developed in a more nuanced way. Urban economies as complex systems are not shaped merely by gradual forms of institutional change, but also possibly by radical changes derived from industries and/or institutions (Gong & Hassink, 2019). Moreover, multiple modes of institutional change are likely to co-exist and co-shape within a local economy; the dynamic mechanisms of regional path development might be more complex than we have understood so far (Martin, 2012). To this end, as put by Grillitsch (2015) and Gertler (2018), it is worthwhile to conduct more multiple case-based comparative and qualitative research on the causal interrelationships among different types of agencies, layers/scales of institutional dynamics and characteristics of path evolution at the urban/regional levels.

Acknowledgements Chun Yang from Hong Kong Baptist University has made great contributions to this chapter. I express our heartfelt thanks to her.

References

Anderies, J.M., Janssen, M.A. & Ostrom, E. (2004). A framework to analyze the robustness of social-ecological systems from an institutional perspective. *Ecology and Society*, 9, 18–36.

Arbuthnott, A., Eriksson, J. & Wincent, J. (2010). When a new industry meets traditional and declining ones: An integrative approach towards dialectics and social movement theory in a model of regional industry emergence processes. *Scandinavian Journal of Management*, 26, 290–308.

Bathelt, H. & Glückler, J. (2014). Institutional change in economic geography. *Progress in Human Geography*, 38, 340–363.

Berman, R., Quinn, C. & Paavola, J. (2012). The role of institutions in the transformation of coping capacity to sustainable adaptive capacity. *Environment Development*, 2, 86–100.

Binz, C., Truffer, B. & Coenen, L. (2016). Path creation as process of resource alignment and anchoring: Industrial formation for on-site water recycling in Beijing. *Economic Geography*, 92, 172–200.

Boas, T.C. (2007). Conceptualizing continuity and change: The composite-standard model of path dependence. *Journal of Theoretical Politics*, 19, 33–54.

Boschma, R. & Frenken., K. (2011). The emerging empirics of evolutionary economic geography. *Journal of Economic Geography*, 11(2), 295–307.

Boschma, R. (2015). Towards an evolutionary perspective on regional resilience. *Regional Studies*, *49*(5), 733–751.

Bristow, G., & Healy, A. (2014). Regional resilience: An agency perspective. *Regional Studies*, *48*(5), 923–935.

China Economic Net. (2014). Taierzhuang created an output of 1.9 billion RMB in the cultural tourism industry over the past thirteen years. [In Chinese]. Available at: www.ce.cn/culture/gd/201401/20/t20140120_2162222.shtml

China Energy News. (2017). Fuxin's wind power: The 'new entrepreneurial' path. [In Chinese]. Available at: http://paper.people.com.cn/zgnyb/html/2017–03/20/content_1759564.htm

Dawley, S. (2014). Creating new paths? Offshore wind, policy activism, and peripheral region development. *Economic Geography, 90*(1), 91–112.

Economist. (2014). The north-east: back in the cold. *The Economist.*

Evenhuis, E. (2017a). Institutional change in cities and regions: A path dependency approach. *Cambridge Journal of Regions, Economy and Society*, *10*, 509–526.

Evenhuis, E. (2017b). New directions in researching regional economic resilience and adaptation. *Geography Compass*, *11*(11), e12333.

Fuxin Statistical Bureau (FSB). (2012). Fuxin Statistical Yearbook. Beijing: China Statistical Press.

Fuxin Statistical Bureau (FSB). (2014). Fuxin Statistical Yearbook. Beijing: China Statistical Press.

Fuxin Statistical Bureau (FSB). (2017). Fuxin Statistical Yearbook. Beijing: China Statistical Press.

Geels, F.W. (2004). From sectoral systems of innovation to socio-technical systems: Insights about dynamics and change from sociology and institutional theory. *Research Policy*, *33*, 897–920.

Gertler, M.S. (2018). Institutions, geography, and economic life. In: Gordon, C.L., Feldman, M.P., Gertler, M.S., et al. (eds). *The New Oxford Handbook of Economic Geography*. Oxford: Oxford University Press, pp, 230–244.

Gherhes, C., Vorley, T. & Williams, N. (2018). Entrepreneurship and local economic resilience: The impact of institutional hysteresis in peripheral places. *Small Business Economics*, *51*, 577–590.

Giannakis, E., & Bruggeman, A. (2017a). Determinants of regional resilience to economic crisis: A European perspective. *European Planning Studies*, *25*(8), 1394–1415

Goldstein, B.E., Wessells, A.T., Lejano, R. et al. (2013). Narrating resilience: Transforming urban system through collaborative storytelling. *Urban Studies*, *52*, 1285–1303.

Gong, H., & Hassink, R. (2019). Co-evolution in contemporary economic geography: Towards a theoretical framework. *Regional Studies*, *53*(9), 1344–1355.

Grabher, G. (1993). The weakness of strong ties: The lock-in of regional development in the Ruhr Area. In: Grabher, R. (ed.) *The Embedded Firm: On the Socioeconomics of Industrial Networks.* London: Routledge, pp. 255–277.

Grillitsch, M. (2015). Institutional layers, connectedness and change: Implications for economic evolution in regions. *European Planning Studies, 23*, 2099–2124.

Hassink, R. (2010). Regional resilience: A promising concept to explain differences in regional economic adaptability? *Cambridge Journal of Regions Economy and Society, 3*(1), 45–58.

Henning, M., Stam, E. & Wenting, R. (2013). Path dependence research in regional economic development: Cacophony or knowledge accumulation? *Regional Studies, 47*, 1348–1362.

Hu, X. & Hassink, R. (2015). Overcoming the dualism between adaptation and adaptability in regional economic resilience. *Papers in Evolutionary Economic Geography, 15*(33), 1–40.

Hu, X., & Hassink, R. (2017a). Exploring adaptation and adaptability in uneven economic resilience: A tale of two Chinese mining regions. *Cambridge Journal of Regions Economy and Society, 10*(3), 527–541.

Hu, X., & Hassink, R. (2017b). New perspective on restructuring of old industrial areas in China: A critical review and research agenda. *Chinese Geographical Science, 27*(1), 110–122.

Hu, X. & Hassink, R. (2017c). Place leadership with Chinese characteristics? A case study of the Zaozhuang coal-mining region in transition. *Regional Studies, 51*, 224–234.

Hu, X. & Yang, C. (2018). Building a role model for rust belt cities? Fuxin's economic revitalization in question. *Cities, 72*, 245–251.

Isaksen, A. (2015). Industrial development in thin regions: Trapped in path extension? *Journal of Economic Geography, 15*, 585–600.

Isaksen, A., Tödtling, F. & Trippl, M. (2018). Innovation policies for regional structural change: Combining actor-based and system-based strategies. In: Isaksen, A., Martin, R. & Trippl, M. (eds) New Avenues for Regional Innovation Systems: Theoretical Advances, Empirical Cases and Policy Lessons. Cham, Switzerland: Springer, pp. 221–238.

Lang, T. (2012). How do cities and regions adapt to socio-economic crisis? Towards an institutionalist approach to urban and regional resilience. *Raumforsch Raumord, 70*, 285–291.

MacKinnon, D., Cumbers, A., Pike, A., et al. (2009). Evolution in economic geography: Institutions, political economy, and adaptation. *Economic Geography, 85*, 129–150.

MacKinnon, D. (2017). Labour branching, redundancy and livelihoods: Towards a more socialized conception of adaptation in evolutionary economic geography. *Geoforum, 79*, 70–80.

MacKinnon, D., Dawley, S., Steen, M., et al. (2019). Path creation, global production networks and regional development: A comparative international analysis of the offshore wind sector. *Progress in Planning, 130*, 1–32.

MacKinnon, D. & Derickson , K. D. (2013) From resilience to resourcefulness: A critique of resilience policy and activism. *Progress in Human Geography, 37*, 253–270.

Mahoney, J. & Thelen, K. (2010). A theory of gradual institutional change. In: Mahoney, J. & Thelen, K. (eds) *Explaining Institutional Change: Ambiguity, Agency, and Power.* New York: Cambridge University Press, pp. 1–37.

Martin, R. (2000). Institutional approaches in economic geography. In: Sheppard, E. & Barnes, T. (eds). *A Companion to Economic Geography*. Oxford: Blackwell, pp, 77–94.

Martin, R. (2010). Roepke lecture in economic geography — rethinking regional path dependence: Beyond lock-in to evolution. *Economic Geography, 86*, 1–27.

Martin, R. (2012). Regional economic resilience, hysteresis and recessionary shocks. *Journal of Economic Geography, 12*(1), 1–32.

Martin, R., & Sunley, P. (2015). On the notion of regional economic resilience: Conceptualization and explanation. *Journal of Economic Geography, 15*(1), 1–42.

Neffke, F., Hartog, M., Boschma, R., et al. (2018). Agents of structural change: The role of firms and entrepreneurs in regional diversification. *Economic Geography, 94*(1), 23–48.

North, D.C. (1990). Institutions, Institutional Change and Economic Performance. Cambridge: Cambridge University Press.

Pike, A., Dawley, S. & Tomaney, J. (2010). Resilience, adaptation and adaptability. *Cambridge Journal of Regions, Economy and Society, 3*, 59–70.

Simmie, J., & Martin, R. (2010). The economic resilience of regions: Towards an evolutionary approach. *Cambridge Journal of Eegions, Economy and Society, 3*(1), 27–43.

Sotarauta, M. & Suvinen, N. (2018). Institutional Agency and Path Creation. In: Isaksen, A., Martin, R. & Trippl, M. (eds) *New avenues for regional innovation systems – Theoretical advances, empirical cases and policy lessons*. Cham: Springer, pp. 85–104.

Strambach, S. & Halkier, H. (2013). Reconceptualising change: Path dependency, path plasticity and knowledge combination. *Zeitschrift für Wirtschaftsgeographie, 57*, 1–14.

Streeck, W. & Thelen, K. (2005). Beyond Continuity: Institutional Change in Advanced Political Economies. Oxford: Oxford University Press.

Steen, M. & Hansen, G.H. (2018). Barriers to path creation: The case of offshore wind power in Norway. *Economic Geography, 94*(2), 188–210.

State Council. (2013). Sustainable Development Planning on Resource-based Cities in China (2013–2020) (in Chinese). Available online at: www.gov.cn/zwgk/2013-12/03/content_2540070.htm

Sjöstedt, M. (2014). Resilience revisited: Taking institutional theory seriously. *Ecology and Society*, 20, 23–31.

Thelen, K. (2003). How institutions evolve: Insights from comparative-historical analysis. In: Mahoney, J. & Rueschemeyer, D. (eds). *Comparative Historical Analysis in the Social Sciences*. New York: Cambridge University Press, pp, 208–240.

Trippl, M. & Isaksen, A. (2016). Path development in different regional innovation systems: A conceptual analysis. In: Parrilli, M.D., Fitjar, R.D. & Rodriguez-Pose, A. (eds). *Innovation Drivers and Regional Innovation Strategies*. London: Routledge, pp, 82–100.

Trippl. M., Grillitsch, M. & Isaksen., A. (2018). Exogenous sources of regional industrial change: Attraction and absorption of non-local knowledge for new path development. *Progress in Human Geography, 42*(5), 687–705.

Van der Heijden, J. (2010). A short history of studying incremental institutional change: Does explaining institutional change provide any new explanations? *Regulations and Governance*, 4, 230–243.

Van der Heijden, J. (2011). Institutional layering: A review of the use of the concept. *Politics, 31*, 9–18.

Yang, C. (2009). Strategic coupling of regional development in global production networks: Redistribution of Taiwanese personal computer investment from the Pearl River Delta to the Yangtze River Delta, China. *Regional Studies, 43*, 385–407.

Yamamoto, D. (2011). Regional resilience: Prospects for regional development research. *Geography Compass, 5*(10), 723–736.

Zukauskaite, E., Trippl, M. & Plechero, M. (2017). Institutional thickness revisited. *Economic Geography, 93*, 325–345.

Zaozhuang Education Council (ZEC). (2014). Notification on the establishment of Zaozhuang tourism vocational education council, Policy document no. 89. [In Chinese].

Zaozhuang Statistical Bureau (ZSB). (2017). Zaozhuang Statistical Yearbook. Beijing: China Statistical Press.

Chapter 8
Conclusions and Research Agendas

8.1 Key Findings of the Book

Throughout the whole book the core aim is to scrutinize the geographies of regional economic resilience of Chinese OICs under different types of crises, and by using multiple perspectives and approaches.

This book contributes to the answers for the proposed research questions in Chap. 1, as follows:

- ***Research Question 1: How is regional economic resilience conceptualized? What are the impact factors affecting regional economic resilience including the nature of various crises/shocks?***

First of all, Chap. 6 developed a conceptual framework that deconstructs the dualism between adaptation and adaptability. It redefined the notion of adaptation and adaptability in a co-evolving and dialectical way, and suggested a typology of relationships between the two. This framework well reflects on many possible ways in which structure and agency from different geographical scales interact for building up place-specific processes, capabilities and trajectories of economic resilience. I thus emphasize that:

Adaptation is an on-going and never-ending process, by which a regional economic system responses to a succession of challenges and disturbances, in order to become fitted to varying environment. It refers to both the current function and performance of being maintained, and the potentials for regional dynamics.

Adaptability is the ability to foster new (or/and renew old) actors, institutions, networks and resources in a regional economy. For a long-term perspective, it is a result of adaptation in which some actors purposively adapt to environments by searching for alternative ways of doing things. Adaptability in some conditions can lead to a new or upgraded form of adaptation.

In terms of the typology, I argue that varieties of the relationships between adaptation and adaptability are indicative of geographically uneven resilience in a long-run

X. Hu, *Uneven Economic Resilience of Old Industrial Cities in China*,
Urban Sustainability, https://doi.org/10.1007/978-981-99-9279-9_8

perspective. It is further illustrated that a combination of multi-scalar contextual and structural impact factors can lead to a certain form of adaptation-adaptability relationship that makes a region's resilience specific.

Secondly, this comparative study also reminds us the importance of regionally differentiated ***institutions and their embedded agency*** in understanding the geographies of OICs. Much work has been done, by using the concept of 'varieties of capitalism' (VOC), to articulate why the evolutionary nature of OICs differs from nation to nation (e.g., Birch et al., 2010). My research, however, highlights that there are varieties of regional models within China. These models are shaped by regionally varied institutional settings. In this chapter, the question of why Zaozhuang is more resilient than Fuxin, is indeed the question of why the institutional environment (including micro niches and macro contexts) differs between the two regions.

Moreover, Chaps. 6 and 7's emphasis on the role of human agency and institutions are actually very much related to the question of why individuals or groups of individuals in some cities are more capable to form or change system-level (or higher yet significant levels) agency closing linking to the core of resilience while others are not. This question is, also, about the role of geographically differentiated institutions and their ways, motives, and targets of change in moulding the relationships between adaptation and adaptability in regional economic resilience. In fact, the whole empirical part highlights that institutions play a consistent role in coordinating the motivations and behaviors of firms, states and people for old and/or emerging industry path development. Therefore, the contributions to ***IEG and agency*** can be summarized as:

(1) the study provides new evidence to argue that there is a need to take an IEG perspective when doing research on China's OICs;
(2) Institutions seem to vary from region to region, despite being under one identical national context; They are under-researched mediators and bridges to understand the relationships between agency, structure and their dynamics, motives and directionalities of change in continuously forming and shaping regional economic resilience
(3) A multi-scalar EEG perspective as a compliment to IEG is called for. Because it can help to clarity what and why impact factors are spatially specific and interactively causal, and what are the micro-level factors from and how they collectively and transformatively generate/impact meso- and macro-level elements and institutions that logically affect the nature of regional economic resilience.

Third, based on our conceptual (Chap. 2) and empirical work (Chap. 5) concerning regional economic resilience the context of (post)-COVID-19 pandemic crisis, we conclude that:

(1) Pandemics are no longer geographically bound within cities. Cities are connected through flows of people and commodities, telecommunications, and supply chains;
(2) The pandemic has thus created a tele network transmission effect with global repercussions via human-to-human geographical proximity and gathering. It has

significantly impacted labor-intensive and service-based sectors and also led to the emergence of new networks and structures of economic activity, such as online working, virtual meetings, and telemedicine. OICs with often weak and single-industry-based structures are particularly exposed to pandemics, whose resilience is the key for future's research not only in terms of conceptualization and empirics;

(3) A city's resilience should be embedded in urban networks to build a global governance system. This involves to adopt multiple perspectives and put relational thinking as the core to the understanding of regional economic resilience. In particular, under COVID-19, resilience is not merely about structure and human agency working together in the economy, but also involved with the supply chain system of production and living goods, medical resource allocation networks, effective social assistance, and social capabilities. This book thus calls for a relational perspective in which global value chain resilience, industry chain resilience and supply chain resilience, as well as social provision chain resilience are all key elements and impact factors for economic resilience in the COVID-19 context.

- ***Research Question 2: How do local economies of OICs in China evolve and adapt with crises, and to what extent are they able (or not) to change old industries and create new ones? And how and to what extent regional path development is linked to and give impact on regional economic resilience for Chinese OICs?***

First of all, local actors in China's OICs tend to regard crises/challenges as context inputs. In Chinese philosophy, for instance, crisis (危机) is a neutral term. It refers to a mixture context with dangers (危险) and opportunities (机遇). Therefore, the question of how local economies adapt to crises, in a Chinese context, is firstly about the question of how local actors read and interpret changing contexts. For example, in Fuxin, local key actors viewed the crisis of the local coal exhaustion as a danger or a problem. All strategies and actions were oriented to fix it, but without awareness to see the crisis as a 'chance' for change (see details in Chap. 6). On the contrary, as presented in Chap. 7, local actors in Zaozhuang not only interpreted the crisis as a 'starting point' for industrial diversification, but also positively contextualized it as an advantage into new policy strategies.

Of course, changing contexts, such as global economic crises, local resource depletion, external market changes, national/provincial-level policy and geo-political policy dynamics, as well the very recent COVID-19 pandemic crisis (as I defined them as contextual factors in Chap. 2) are part of impact factors affecting the agentic ways and the speed/duration of adaptation in OICs—which leading to regionally uneven economic adaptability and resilience (also see Chap. 6). Besides the contextual contexts, structural factors play a fundamental role in affecting and making the economic resilience of Chinese OICs.

In addition, I found out that the 'positionality' of an OIC in a specific and history-conditioned local-central relation strongly affects the motivation of local actors.

Chinese local officials and SOE leaders are not only embedded in local institutional contexts, but also, to a large extent, are affected and coordinated by top-down political and policy forces. This can be exemplified by the analyses in Chaps. 4, 6 and 7. It has shown that local stakeholders in both regions shared a strong awareness of the political relationship between the central state and the locality. They knew very well of what they could do, what they could not do, according to the macro institutional structures. For instance, since Fuxin's economy for a long run has been predominated by upper governments, there is a recognized institution that the power scope and legitimacy for the local state is limited. Such institution is well accepted by local officials, allowing them only to depend on the top-down policy interventions when it comes to crises (Chap. 7). The close local-central relation between Beijing and Fuxin, which is highlighted in Chaps. 6 and 7, basically undermines the capacity and creativity of local stakeholders to foster industrial change and enhance regional economic. This relation only put them in a narrow position that they tended to welcome and implement all sorts of national poli-cy actions. There is thus a lack of concern among Fuxin's actors about whether national policies were suitable to local-specific conditions.

Secondly, in general, China's OICs are more likely to evolve in a cumulative process. The concept of path dependence can partially explain the path continuity of the coal-mining industry in the case regions. Basically, due to the coal-based structure of China's energy production and consumption, the industry will continually keep growing. The deregulation of cross-regional mergers and acquisitions (M&As) for coal SOEs well manifests the robustness of this industry. From this book, I found that the coal industry in both case regions followed a strong path-dependent way of development. Although new growth industries emerged, they were, indeed, strategically facilitated without harming the coal industry. The empirical evidence suggests that China's OICs are not in a hurry to radically transform their economies. Simply because the energy-based economic structures are not, and will not be, meant to be completely changed due to multi-scalar and structural factors. It is expected that in the future many China's OICs may diversify over time, but the mining industry, and other energy industries, will remain growing, and will last for a long-time period.

In terms of new path creation, we observed that no new industry is created from scratch in both case regions. Either (pre-) existing industrial legacies and assets or uncommitted local resources are more or less exploited and utilized to support new path creation. For instance, the coal-chemical industry in Zaozhuang was initially built upon the pre-existing production base of fertilizer. Likewise, the tourism industry in Zaozhuang was selected not merely because of the new mayor implementing the national policies of developing 'Red Tourism' locally (Hu & Hassink, 2017). It is also because of the rich historical tourism-related assets in the region. Even the wind power industry in Fuxin, which seemed to be detached from the local industrial system, was settled due basically to the rich local wind resources (Hu & Yang, 2018). Therefore, it is concluded that China's OICs are more likely to evolve by drawing upon pre-existing assets. I argue that no matter how new paths are fostered (in an endogenously or exogenously way), place-based history and its embedded assets, either in various forms and actors, often sets a scope or provide initial conditions for new paths to emerge and old paths to adapt—namely the nature of resilience strategies and actions.

Thirdly, at least from the empirical evidence provided in this book, I argue that economic resilience of OICs in China is, basically, about the ability of the local government and its constitutive officials, as well as its supported firm and industry actors, to foster industrial change and evolution during crises. There are several impact factors identified in the book that enabled regional economic resilience: (1). As shown in Chaps. 6 and 7, the replacement of local state leadership may result in a high likelihood of local institutionalization for new industry formation. Moreover, Chap. 7 stated that 'airborne cadres', compared to other types of officials, are incentivized by upper authorities to 'make changes' in new localities. This means that OICs with airborne cadres might have a higher potential of economic resilience than OICs with 'locally rooted cadres'; (2). Economic resilience is contingent on the ability of local state actors to read and interpret the nature and features of crises, and to purposely mobilize multi-scalar actors and assets, as well as to link local conditions to macro policy agendas, for industrial adaptation and development. In other words, resilience in China's OICs is very much related to the sensitivity and resourcefulness of local officials, in the forms of state governance actions, abilities, styles, and directionalities (as mainly addressed in Chaps. 4 and 5).

However, there are also some negative factors that make local officials both cognitively and behaviorally constrained, when it comes to high-quality and long-term economic resilience. Chapter 7, for instance, clearly demonstrated that the Chinese hierarchical SOE ranking system in OICs—namely the co-existence of SOEs with different ownership in one region—gave rise to the incoherent form of local institutional structure. This incoherence thus led to a situation of inner-regional power segmentation and conflicts. And it prevented local officials to fully and efficiently manage the whole local economic system. All economic governances of the local state must make sure that there would be no harm to the vested interests of higher-hierarchy institutional bodies in the city.

- ***Research Question 3: why do Chinese OICs adapt differently to certain shocks on the ground and how to understand their unevenness in regional economic resilience in terms of process, mechanism and effectiveness on the ground?***

Drawing on the comparative case study in this Chapter, I concluded that, firstly, the adaptation in both regions underwent a solid old path continuity and extension process. This identical adaptation was mainly enabled by both the top-down national policy and local inherent need for development. However, the adaptability of the two regions evolved in a very divergent way:

Zaozhuang City: new industries (e.g., the tourism industry) were fostered and op-erated mainly through local actors mobilizing local assets. In other words, the adaptability in Zaozhuang is derived from the adaptation process in an endogenous way. Moreover, adaptability seemed to have a positive impact on adaptation. The rise of the tourism industry changed regional images, which facilitated the

investment niche and in turn helped the business diversification of local coal firms. It is proven that adaptability and adaptation mutually benefited from each other and, had a complementary relationship
Fuxin City: new industries (e.g., the wind power industry) were planned and managed by extra-local actors. The long-term centrally planned economy in Fuxin has led to a severe lack of bottom-up economic accountability for local actors. This planned history, however, still plays an important role in Fuxin, partly because of the mentality of local actors relying on top-down helps, and partly because of the central state itself recently trying to reshape the industrial landscape of northeast China. Thus, the evidence tells that the adaptability in Fuxin is mostly detached from the local coal adaptation system. Adaptability was developed by national-state forces in an exogenous way. The adaptation and adaptability are not related, and evolve in parallel. Since the newly emerged industries are operated by policy-driven external actors, it is expected that such adaptability might undermine the adaptation process from a long-term perspective, when it comes to national leadership replacement or macro national policy shifts.

Secondly, Chap. 7 steps further based on the conceptualizations and empirical understandings in adaptation and adaptability in regional economic resilience. It adopts a more nuanced perspective focusing on the causal relationships between ways of micro-level agentic dynamics, modes of institutional change and types of industry path development.

- What we have found is that during the past decade, ***Zaozhuang's resilience*** is characterized by positive layering and conversion. Positive layering refers to the addition of new layers for a cognitive shift among local agents. And conversion relates to the revisions on the prevailing GDP-ism for new goals. Under this mode of institutional change, Zaozhuang's economy has undergone a positive restructuring, namely renewal and related diversification. Such restructuring is driven by the purposive actions of local state elites taking advantage of institutional flexibility, and making strategic combination of exogenous policy opportunities and endogenous institutional assets for a far-reaching goal. This underpins the resilience with competitive dynamics for long-term development.
- ***Fuxin's experience*** is quite different. All local responses have centered on two aspects: maintaining the existing coal-based economic institution, and keeping in line with the national policies for sustaining the close local-central relationship inherited from, and cumulatively enforced by, the previous planned economy. This leads to institutional thickening that constrains restructuring (Martin, 2010). Moreover, although the national state has attempted, by deploying new sectors, to help Fuxin to transform, little attention has been paid to local institutional conversion. And the new industry controlled by extra-local agencies as new layers only make the existing institution thickened while lacking interaction with the coal industry. Fuxin's economy thus features a self-reinforcing industrial adjustment, making its resilience with fragmented processes, limited potentials and transient power for industrial change at present and the future.

Our empirical analyses suggest that in the Chinese OIC context the state and SOEs at multiple levels play salient roles in affecting dynamics and geographies of institutions and resilience (Hu, 2014, 2017; Hu & Yang, 2019). Some cities are likely to practice institutional ambiguities because of the absence of national 'hands', and thus are able to facilitate positive layering and conversion. While others may fail to take initiatives from the bottom, and their dynamics tend to be at the mercy of, limited by, top-down policy interference (Hudson, 2005). We contend that the ways in which institutions evolve might be contingent upon how agents read and interpret their situated positions of local-central geometrics, as well as on how they adapt to local and broader changing political economies (Boschma & Capone, 2015; Raco & Street, 2012). Resilience of cities can be in a way understood as geographically divergent ways in which actors enable or constrain institutional dynamics. It can, at base, be seen as processes of regional-specific potential/advantage and institutional dynamic power which matter for long-term economic development (Martin & Sunley, 2015).

Secondly, in general, China's OICs are more likely to evolve in a cumulative process. The concept of path dependence can partially explain the path continuity of the coal-mining industry in the case regions. Basically, due to the coal-based structure of China's energy production and consumption, the industry will continually keep growing. The deregulation of cross-regional mergers and acquisitions (M&As) for coal SOEs well manifests the robustness of this industry. Based on our empirical case studies, I found that the coal industry in both case regions followed a strong path-dependent way of development. Although new growth industries emerged, they were, indeed, strategically facilitated without harming the coal industry. The empirical evidence suggests that China's OICs are not in a hurry to radically transform their economies. Simply because the energy-based economic structures are not, and will not be, meant to be completely changed due to multi-scalar and structural factors. It is expected that in the future many China's OICs may diversify over time, but the mining industry, and other energy industries, will remain growing, and will last for a long-time period (Hu, 2017).

8.2 Contributions to Economic Geography

8.2.1 Contributions to the Literature of the Restructuring and Resilience of OICs

First and foremost, my study applied western modern concepts and theories, such as EEG, IEG and GPE, in a non-western OIC context. The existing literature on the restructuring of OICs has been intended to focus on the role of firms and industries, and within a strong single-case study setting. However, this book focuses more on how multi-scalar agencies and institutions play roles in affecting adaptation and adaptability of OICs in the Chinese authoritarian context. The Chinese experiences provide some refreshing findings and complementary perspectives to the existing literature about how OICs evolve and adapt over time in various settings of crises contexts.

Secondly, many scholars suggested that the role of institutions should be more explored when analyzing the evolution of OICs (Dawley, 2013; Hassink et al., 2019; Hudson, 2005; Morgan, 2012). My research has mainly contributed to this direction. Chapter 5 highlighted the role of hierarchical political economies among different SOEs and states in leading to a specific mechanism of path creation. This mechanism differs from what the existing literature usually tells. Furthermore, Chap. 6 explored how multi-scalar agency interact with structure with different modes of institutional change that are devoted to certain patterns of industry path development in OICs. On the one hand, it shows how powerful and efficient the Chinese states are in terms of fostering new growth paths in OICs. On the other hand, it states that every state-led path creation occurs with reasons. The history of inter-state interaction and relational activities can explain why new paths emerge in certain cities, and why certain paths but not others are selected to emerge.

Thirdly, Chap. 6 contributes to the understanding of the role of human agency in affecting the restructuring of OICs. It systematically articulated how Chinese local state elites consciously managed to cultivate and facilitate new path creation in an OIC, by critically using the notion of place leadership. The evidence from Zaozhuang rejects the traditional perception, that in an authoritarian context local leadership is inefficient and does not lead to innovative governance on economic resilience (Chap. 7). I argue that Chinese local state leadership does not simply refer to some sort of formal and conceive power to get things done. Local institutionalization in China's OICs is indeed fostered by both in-formal and formal leadership. Interpretative leadership, for instance, is necessary for newly appointed local leaders ('airborne cadres'-空降领导), if they attempt to use formal power efficiently, and if they want to change traditional institutions for enhancing and preparing long-term economic resilience (Hu & Hassink, 2017). This contribution to the OIC literature is significant. It provides a new perspective of how micro-level dynamics—particularly human-based individuals (groups of individuals)—matter for local institutionalization and governance innovation for endogenous industrial change (Hu & Yang, 2019).

8.2.2 Theoretical/Paradigmatic Contributions to EG

(1) *Path dependence and lock-ins*

First, the concept of path dependence has shown explanatory power in the Chinese OIC context. In the Chinese context, the effects of path dependence are mainly derived from the long history of economic and political centralization. These effects have penetrated into local firm institutional networks and bureaucratic systems. The Chinese experiences, to a large extent, go beyond the effects of 'place dependence'.

Because, both in Zaozhuang and Fuxin, the extension of the old mining path is not only enabled by local specific forces, but more affected by trans-local forces related to macro policy changes.

Secondly, the concept of lock-ins, which has been extensively used in explaining the de-cline and stagnation of OICs in the literature, is partly applicable to the Chinese cases. The three types of lock-ins (namely functional, cognitive and political lock-in) can be observed both in the Zaozhuang and Fuxin region. The SOE-dominated and highly specialized economies have led to a high degree of functional lock-in. More importantly, the reason why the coal industry is maintained and extended is that local officials were both cognitively and politically locked into the coal-driven political economy. The so-called 'GDPism' pressured local actors to firmly protect their 'money tree'—the mining industry. Not only the industry contributes most to local economic benefits, but also it fundamentally matters the career prospect for each individual local official. Therefore, the lock-in concept seems useful and meaningful to explain the path-dependent feature of regional adaptation in Chinese OICs.

However, the Chinese evidence critically shows that the effects of lock-ins are not merely restricted at the firm level or local social/political context as Grabher (1993) suggested. The lock-in concept needs to be put into a multi-scalar and non-firm centric context. For Chinese OICs, lock-in effects are not only locally built, but, basically, are conditioned by multi-scalar institutions, politics and economies. Moreover, the findings support Martin's (2010) dynamic idea on lock-in. lock-in is not a static stage that follows the law of 'once locked-in, never get out'. It evolves with changing contexts. For example, the lock-in effect on the coal-mining industry in both case regions were reinforced by new national polices. The other way around, lock-in does not necessarily mean that local actors who are firmly locked in certain old industries have no (or few) potential to foster new industries. The lock-in situation in some circumstances can be turned into positive sources for industrial renewal and diversification (Hassink, 2010; Hassink et al., 2019).

(2) *Path creation, path plasticity, and path development*

To examine the uneven economic resilience of Chinese OICs, the book (Chaps. 3–7) puts quite a lot of efforts on exploring the processes, characteristics and forms of path development (including old path dynamics and new path creations). It is proven that new paths can be largely built upon extra-local assets, not necessarily following a place-dependent, endogenous and branching process. The results have shown that an endogenous process of industrial diversification—namely local existing related assets are used to enable new path formation—can be undermined and interpreted by local incoherent political economies. My key argument is that the Chinese administrative ranking system among SOEs and states in a regional economy has led to a high heterogeneity of agents, which might distort the rational ways for new path creation.

Thus, I argue the firm-centric endogenous EEG approach, in terms of the concept of related variety and regional branching, is not sufficient to explain the complexity of path creation in Chinese OIC contexts. In my research, institutions, policies and politics both from local and extra-local spatial levels are the keys to thoroughly understand the processes of regional path creations in China.

An incomplete form of power decentralization and marketization gives rise to varieties of agents and power incoherence within an OIC. On the one hand, planned economies and top-down institutions and, even their legacies stay firm, and continually frustrate the change of regional economies. On the other hand, local stakeholders are politically incentivized to do something new on economic development. In this context, the ways of path creation are largely related to how local elites balance the power and interest in-between.

In this sense, the concept of ***path plasticity*** seems applicable to analyze and explain the Chinese-specific stories. As Strambach and Halkier (2013) stated, "there is always a degree of plasticity which enables also that innovation with a minor degree of complementarity within the well-established institutional setting of paths may come into being" (p. 1), and "[there is] always a limited range of choice of a well-established setting in creative ways for the development of innovation without breaking out of the path" (p. 1). In the case studies (e.g., Chaps. 6 and 7) confirm these arguments. The evidence further suggests that (new) path development is about many possible combinations—mediated by agency, through its work on institutions and structures—continuity (via adaptation with path dependence effect) and change (via adaptability with path innovation effect). Resilience is reflected by the process of path development, in which adaptation and adaptability interplay and form the nature (including feature, ability, capability, and speed/motive of action, etc.) of how and why city economies respond and adapt to crises for whom.

Here, in the Chinese OIC context, economic resilience is very much about the resourcefulness of the local state and its constitutive state elites and institutions (Hu et al., 2020). Path creation in a Chinese OIC context must be, more or less, a result of power and interest coordination among varied stakeholders from multi-level geographical and administrative spaces. This coordination can be seen as an achievement that local actors maintain (or at least no harm to) old institutions and simultaneously facilitate new industries for their own interests. This is why in both empirical regions old path extension and new path creation can co-exist harmoniously, but, not in a mutually cohesive and inclusive way. Therefore, I think the idea of path plasticity is potentially useful to explain why Chinese OICs diversify over time, but still with a slow and problematic restructuring process.

At last, the evidence also suggests that regional path creation does not always lead to a positive and long-lasting impact on regional development. New paths might be created on temporary purpose, and thus may exist for only a short period of time (Hu & Yang, 2018). This has been evidenced by the creation of the agriculture and food industry in Fuxin in Chap. 7. This industry was fostered merely in order to radically provide a low-skill required labor pool for coping with unemployment caused by the declining mining industry. Therefore, I argue that the question of 'new path creation for what?' (Pike et al., 2010) needs to be paid more attention to in future path development research. There is a need to keep the term of 'new path creation' conceptually neutral, and so as to the notion of new path development—as any 'creation' is oriented to and/or cannot escape from the process of 'development'.

(3) *Adaptation, adaptability and regional economic resilience*

As many newly introduced terms in EG, adaptation, adaptability and regional resilience are still fuzzy in definition, and both conceptually and empirically in its infancy. Chapter 2 has tried to conceptualize these terms in a systematic way, and also demonstrated with empirical case studies in Chap. 6. The dualistic idea, inherited from Grabher's traditional static lock-in concept (Grabher, 1993), has been criticized in the chapter. Adaptation and adaptability cannot be discussed or analyzed separately, as they are internally related. They should be understood and conceptualized dialectically. Because adaptation as process indeed incorporates dynamics and sources for adaptability. And adaptability as a changing ability is an integral part of the process of adaptation. They not only co-exist in many contexts, but also co-evolve with each other. Instead of introducing new thoughts from other related social sciences to conceptualize resilience, this book mainly deepens the understanding of this notion by discussing the relationship between adaptation and adaptability. It is also emphasized that regional resilience should be conceptualized and operationalized with an economic geographical perspective. The key contribution in the book is that there are varieties of adaptation-adaptability relationships in regions. Different forms of relationships are highly indicative of the uneven economic resilience of regions. I also argue that regional resilience is not just about capacities and capabilities for a short-term recovery agenda, it is about functionality, transformability and long-term sustainability of regional development. Although this understanding, of course, is not universally applicable, it does provide a new vision to further scrutinize the idea and to make sense of it in future conceptual and empirical work.

(4) *Institutional Economic Geography (IEG), Relational Economic Geography (REG) and Geographical Political Economy (GPE), towards 'engaged pluralism'*

This book, particularly in Chaps. 3 and 6–7, paid much attention to the relations and behind the 'power geometrics' (Yeung, 2005) among states. It is argued that the political distance between the local and the central state is essential for the human agency of economic change in China's OICs. This argument seems not only applicable to OICs, but also to other types of regions within China (Economist, 2015). Therefore, in terms of the contributions to REG, the study moves beyond the focus on the local relations and networks of firms, suggesting that historically informed multi-scalar geographical politics and power relations can be the core to better understand uneven regional development and evolution. It thus further argues that the focus of EG research should shift from relational assets/behaviors within and beyond firms, to their embedded positionality and relationship attributes among multi-scalar political-institutional environment and frameworks (Bristow & Healy, 2014; Martin, 2010).

In the end, as Barnes and Sheppard (2010) argued, 'nothing includes everything', each chapter in the thesis can only partially answer certain specific questions, but cannot fully understand them. As questions themselves are multi-faceted and cannot be explained by one or two concepts and perspectives. Although each chapter draws different theoretical sources to try to deal with the same topic, the collection of work as a whole, in fact, shows a combination of different perspectives and concepts—a sort of 'engaged pluralism' to explore the topic of the restructuring of OICs. Here, I have to confess that doing research on the evolution of China's OICs is challenging because:

A hybrid economy consisting of highly heterogeneous but intertwined actors and institutions needs us to pay more attention to the interactions and relationships between them. These relations are dynamic in form but also are conditioned by the past, consistently shape and are shaped by multi-scalar institutions and power configurations. In this sense, REG, IEG and EEG approaches should not be separated when it comes to analyzing the evolution of OICs.

(1) the EEG approach may, to a large extent, fail to explain the evolution of China's OICs. Due to the planned and centralization history and the uncompleted decentralization, Chinese OICs lack endogenous ability for industrial change. The dynamics for industrial change are more likely to be fostered by exogenous actors. Thus, EEG may have strong explanatory power to address why old industries/paths continue to stay, but may fail to explain why and how new paths emerge and evolve.
(2) In order to better understand the question of geographically uneven regional economic evolution, we need to carefully choose appropriate concepts in EEG according to specific contexts, and integrate them with IEG, REG and GPE ideas, being theoretically 'pluralist'.

8.3 Relational Resilience and Chain Resilience Under the COVID-19 Crisis: Some New Thoughts

Chapters 2, 3 and 5 suggest that the nature of crisis might matter significantly for regional economic resilience. The recent COVID-19 crisis has been proofed to generate highly distinctive impacts and outcomes on regional economies. It brings about geographically uneven challenges and shocks, and involving different actors, organizations, and institutions at highly diverse geographical scales. This crisis is a shock on any forms of socio-economic systems, it often generates chain reactions across space, having the power to change the structure of networks, connections and systems in which economic actors are embedded in (Hu et al., 2022).

As Chap. 2 has proposed, In the context of deepening economic globalization and regionalization, China's contemporary economy is intricately intertwined within global networks such as GPN and GVC, involving supply chains, industrial chains, and value chains. As the international geopolitical and economic landscape grows increasingly complex and challenging, these chains are frequently subjected to systemic disruptions. China's deeply integrated economy is finding it progressively difficult to remain insulated. Confronted with this reality, there is an urgent need to move beyond traditional evolutionary resilience conception and to reevaluate the processes and mechanisms of resilience from a relational perspective of chains

(MacKinnon et al., 2019). Against this backdrop, the concept of "chain resilience" (链韧性) has been increasingly relevant attention to rethink the concept of regional economic resilience under the COVID-19 pandemic. The integration between 'chain resilience' and 'evolutionary resilience' is potentially powerful in the explanation for "regional" (but nationally/globally connected by chains and networks) resilience.

The preceding discussion indicates that there exists a complementary perspective between 'chain/relational resilience' and 'evolutionary resilience.' The research perspectives found in 'chain resilience,' such as multi-geographical scales, multi-element correlations, and external regional linkages, can broaden the theoretical understanding of regional economic resilience, addressing its historical limitations of focusing solely on static regional attributes and endogenous factors—'as structure' (Boschma, 2022). Building upon this notion, I therefore incorporate the 'chain resilience' perspective into the conceptualization of regional economic resilience, aiming to construct a novel analytical framework for regional economic resilience under the COVID-19 context. The goal is to facilitate a theoretical dialogue and integration between these two resilience research paradigms.

Firstly, based on what we found in Chap. 5, The evolutionary perspective of regional economic resilience mainly focuses on the preset role of regional structural characteristics on economic resilience (Boschma, 2015), while the relational perspective in form of chain resilience emphasizes the structural dynamics linking to multi-scalar chains of firms, industries and other assets. Combining the theoretical and empirical research results of these two types of resilience, I argue that regional economic resilience is a complex dynamic process involving multi-geographical scales, multi-subject interactions, and the relational/contextual sensitivity of endogenous and exogenous factors. A region or city's economic resilience is never self-sustained and self-sufficient on its own, but affected and structured by broader relations and networks in the forms of industry, value, supply chains, as well as mediated by distant social ties and political links.

I, therefore, propose a new conceptual framework of regional economic resilience from a relational approach which consists of two analytical objects: *the nature of crisis/shock* and *capabilities of resilience*. For the former, it includes the scope, duration, depth, and object of influence, the mechanism, speed and scale of risk/shock transmission, impact mechanism, as well as potential destructiveness. The notion of capabilities of resilience can be deconstructed from the two dimensions, namely, "structural/agentic factors" and "endogenous/exogenous factors". The basic idea of applying this analysis framework is organized with four steps: (1) by identifying key characteristics and elements on the nature of a certain crisis; (2) resilience' capabilities can be understood from understanding the interplay between structural and agentic factors, with a multi-scalar view focusing both regional and external factors, respectively; (3) interpret the interplays via the views of relations, links and networks associated with the 'chain' concept; and (4) understand the resistance, recovery, adaptation and transformation capabilities of a city to match the shock/crisis nature and the chain dynamics. And then comprehensively evaluate the resilience of the regional economy. To be clear, I elaborate the analytical logic and key elements that are worthy for analysis in the framework as shown in Fig. 8.1.

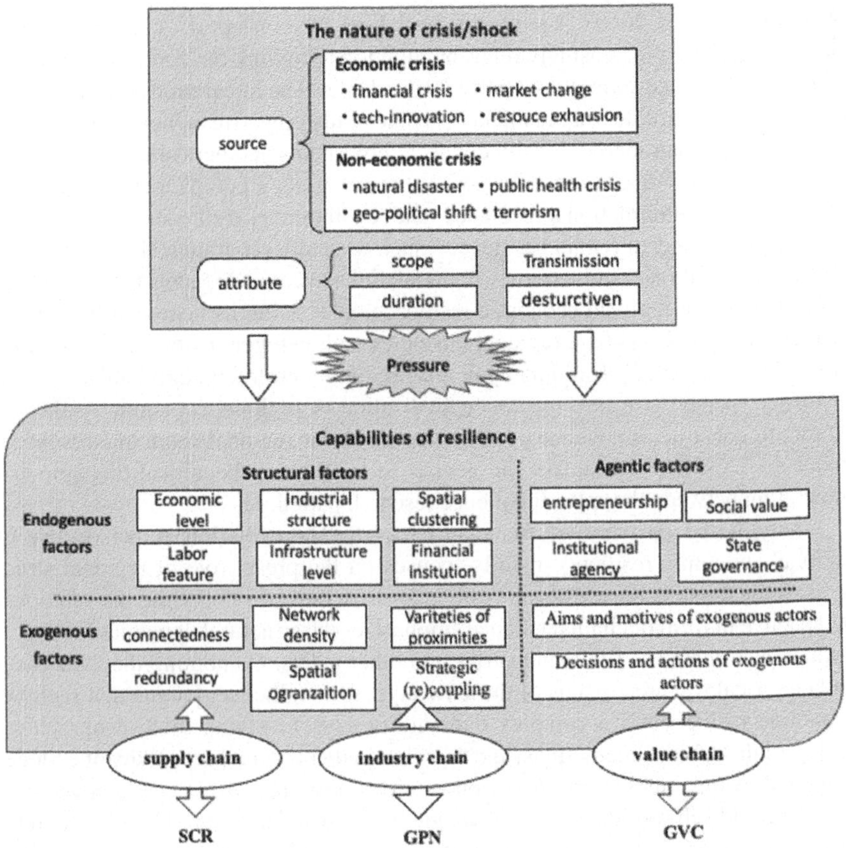

Fig. 8.1 Analysis framework of regional economic resilience from a relational perspective

There are two dimensions for the framework: for one thing, the analysis on the nature of crisis/shock. Recent studies on 'resilience' and regional economic resilience have highlighted the significant influence of crisis characteristics on the continuity and vulnerability of economic activities (Glaeser, 2022). By studying the differences in crises/shocks, there exist noteworthy variations in the determinants and mechanisms contributing to regional economic resilience. There are at least four aspects relating to the nature of crisis/shock in the following: (1) Extent of impact (community-based, localized, regional, national, or global); (2) Duration (short-term, medium-term, or long-term); (3) Mechanisms and geographies of transmission (via what kind of chain, by what kind of object, involving what kind of spatial boundary); (4) Potential level, depth and scope for disruption and damage.

For another, the analysis on the capabilities of resilience. The manifestation of regional economic resilience lies in its ability to effectively withstand risks and shocks. It not only involves the ability of achieving economic recovery, but also a long-term capabilities and power of dynamics of ultimately reaching a new level of development that can surpass the initial state and structure (Gereffi, 2020). These capabilities require on the collaborative interaction of multi-scalar structural and agentic factors working for and within a region/city on the ground (Dodds et al., 2020). Four aspects of analysis in this dimension are:

(1) *Endogenous structural factors* within the region encompass elements such as economic development level, industrial structure, spatial concentration, infrastructure, financial environment, social capital, and labor characteristics. These factors can generally be quantified through economic indicators such as total GDP or per capita GDP, the proportion of each industry's output or value-added, the percentage of fixed asset investment, as well as specialization, related diversification, and unrelated diversification indices.
(2) *Exogenous structural factors* involve attributes related to chain structures, such as the length of supply chains, industry chains, value chains, network density, connectivity, visibility, flexibility, strategic cooperative relationships, information-sharing mechanisms, and spatial layout.
(3) *Endogenous agentic factors* pertain to the behavioral decisions and innovation activities of firm unfolding within a certain city, which generate change agency and dynamics on organization, management, technology, and labor. These factors also encompass entrepreneurial spirit, institutional dynamism, socio-cultural values, and local government governance measures.
(4) *Exogenous agentic factors* encompass the strategic objectives, organizational methods, and behavioral decisions of entities outside the region, including adjustments and innovation actions and impact from large multinational corporations, government agencies, and consumers in the market.

Within this framework of resilience capability analysis, endogenous impact factors constitute the traditional dimensions of regional economic resilience measurement, while endogenous factors, viewed from the relational perspective of 'chain resilience', introduce a new dimension to examine more micro- and meso-level interplay between networked structure and agency in regional economic resilience. The analysis of these four aspects of factors facilitates the integration of an 'evolutionary perspective' with a 'relational perspective,' adding new conceptualization sources for regional economic resilience under the COVID-19 crisis and alike.

8.4 Policy Implications for the Resilience Development of OICs in China

Chinese OICs, particularly in terms of resource-based regions, have already been the core regions supported by national policies. Besides the scientific objects, one important aim of this research is to generate some policy implications. I thus propose three points, which firstly identify crucial deficiencies/problems and then put policy-related comments in regard to economic resilience and future development of China's OICs, as follows:

(1) Throughout the empirical work the most constraining factor of OICs for change, in my view, is the situation of incomplete decentralization. There is a consistent existence of administratively incoherent local economic systems, made up by different ownerships and ranks. Particularly in terms of energy and natural resource sectors, national or provincial-level states still keep them under control for strategic reasons. This can be well evidenced by the coal-mining industry in the case regions. They are mostly owned and operated by the provincial state, rather than the local state. Given this, China's OICs have suffered from a lack of administrative freedom and self-reliance ability for place-based endogenous development.

I argue that the incomplete decentralization of SOEs in OICs caused by the Chinese rigid ranking system strongly prevents the restructuring process of OICs from going smoothly and forward (Hu, 2017). The power games among different bodies of interests, which created and (re)produced by this system, result in unnecessarily high costs. Such costs make local agents limited in vision and action. Therefore, future policies should be geared to the decentralization of SOEs to the locality. The top-down political and administrative 'hands' need to be removed from local/regional economic systems. If the local state of OICs gains enough or loosely defined autonomy on economic development, it is expected that local agents may be able to make more place-based and self-benefiting policies for industrial change (Hu & Yang, 2018, 2019).

Of course, I am not suggesting a radical reform on SOEs in OICs. Because many economies of OICs in China are still relying on, and even surviving on national policy support. A radical change would be harmful to long-term or healthy economic resilience of OICs. I am therefore suggesting an incremental process of decentralization with steady step-by-step and fine-tuned reforms that are not at the expense of both local and extra-local actors, assets and power.

(2) The second point I want to address is about the ways of national policy interventions. I cannot deny the importance and necessity of national preferential policies for China's OICs, particularly in the current post-crisis time. However, the Fuxin case has shown that national policy interventions seem to focus more on deploying 'strategic' new industries mainly with national macro interests. These policy interventions lack concerns with local conditions and demands. In some conditions, new industries are fostered quickly in OICs merely to meet macro industrial planning aims, and may be not well embedded into local economic systems. In my view, such kind of path creation mainly driven by top-down fast policies in OICs is problematic and should be avoided.

It is suggested that the national state should pay more attention to place-specific conditions and needs of OICs. National policy-makers should carefully examine whether top-down policies are able to facilitate the endogenous ability of OICs for industrial localization and relatedness. In my view, the traditional policy idea of 'taking care' or 'helping to fix problems' for less-favored areas needs to revisited. Helping to launch new industries in OICs does not mean to solve long-lasting structural problems in regions, not merely limited to respond to certain crises for short-term resilience. Thus, my key policy advice is that national policies should be inclusive, sustainable and flexible in terms of aims and operationalization in the locality. A good national policy is a policy that can be well integrated to local existing policy and institutional frameworks. It is oriented to long-run industrial structure transformation, rather than temporary new path creation or temporary 'good' records of resilience.

(3) Last but not least, relating to the Chinese cadre cross-region reshuffling system, there is a dilemma that on the one hand newly appointed local leaders are subject to change, while on the other such change may be temporary. Without this system, it may have another dilemma: are locally rooted leaders more able to generate more down-to-ground industrial development policies with a long-term perspective? Or are they also more likely to be locked into certain paths. As proven clearly in Chaps. 6 and 7, Chinese local state leaders are the key actors to make changes in OICs. I think these dilemmas should be regarded as important aspects for future policy considerations. I would suggest the national and provincial state to 'drop' younger, educated 'airborne cadres' in those OICs where regional economies have confronted long-lasting problems and challenges. However, it is also needed for upper states to consider whether it is worthwhile to up-date new ones when the five or ten-year tenure of the presence of local cadres come into the end. If a current 'airborne cadre' performs well, and is dedicated to a long-term incremental transformation rather than a 'temporary' record. Then an extension of his or her stay in existing region should be encouraged.

In general, I claim that China's OICs critically need continued, incremental and long term plans and strategies for sustaining, preparing and making long-term economic resilience, rather than technological breakthroughs or radical reforms or any other fast policy fix or emergence/context-induced problem solving. Basically, these cities' real problems are fundamental, institutional and cross-region 'links, chains and networks' involved, which matter for a broader systematic work involving collective agency by multi-actors at multiple geographical scales. I particularly argue that under the context of (post-) COVID-19, combined with the global grant geo-political shift, EEG's explanatory power and conceptual rigour to the regional resilience research turn to be weak. REG, particularly the chain perspective, and the agency approach in GPN and GPE paradigms, instead, can provide more effective tools to rethink the nature of and to reconceptualize regional economic resilience, not only for OICs but also for other types of cities/regions. In the end, building up a strong regional economic resilience is a systematic work which needs time and patience, neither strong policy tools nor regulation methods, can get sustainable resilience done once and for all. This allows us to give more emphasis on the role of agency and institutional design in form a more grounded, bottom-up dynamics with broader institutional and organizational support and consensus.

8.5 Future Research Agendas

Finally, I propose several near-future research avenues related to regional economic resilience research in China.

First and foremost, in general, future research needs to focus more on the evolution of OICs in emerging economies under different types of crises. The current theoretical accounts are very much centered on, derived from, and limited to Western contexts (mainly Europe and North America). The endogenous model of restructuring may not work in explaining the adaptability and resilience of OICs in emerging economies in the globalized and uncertain context. In my view, for example, China's OICs can be seen as 'battle grounds', where old style planned institutions and new market-based economies co-exist, and both endogenous and exogenous forces are mixed in a complex way. A combination of path dependence and creation, or put it in other words, a situation of divergent multiple paths of development in OICs provides promising 'fertile grounds' for testing and theorizing back on EEG in particular, and EG in general.

Secondly, the role of the state needs to be placed at the core for future OIC resilience research. The question of how people in the state do is poorly understood and also under-theorized in the OIC literature. Particularly in an authoritarian context, local officials, rather than grass-rooted entrepreneurs, are actual decision-makers, business operators, policy practitioners and institutional designers. Within this context, how economies of OICs are restructured is largely dependent on the visions and actions of local officials. I suggest that the notion of place-based leadership and institutional entrepreneurship might be useful to the analysis on the topic. Regarding to the Chinese context, future research questions can be: (1) how do local officials scan multi-scalar contexts and how do they act accordingly? And why do some OICs put top-down policies in action more quickly than others? (2) What are the differences of regional economic resilience between OICs with 'airborne cadres' and OICs without? (3) How do multi-scalar institutions play a role in affecting the motives and behaviors of human agents on regional institutional change and industrial path development? And how are modes of institutions, patterns of organizations and socio-economic structures in the multi-scalar links, chains and networks established and molded by change agents of resilience, for what and for whom?

In the end, to fully understand the unevenness of economic resilience in OICs, an engaged pluralist approach is needed. Research objects including enterprises, states and human actors, among others, need to be put into a relational, evolutionary and institutional context. Firm actions are not isolated from their relational assets. They are also embedded in multi-scalar institutions, often in forms of socio-economic links, chains and networks, and consistently performing with more or less political meanings (Hassink et al., 2019). Precisely, firms and states themselves do not act; it is the people within them who do act. At the micro level, it is people who consciously maintain and change rules and institutions, who do things that are impacted by and also shape meso- and macro-level structures and systems of relations, business networks and political-institutional frameworks. Future research thus should combine with various theoretical approaches particularly in terms of REG, GPN, GVC, GSC and GPE in a complementary way to ask more thorough questions within (E)EG.

Methodologically, cross-national comparison with multiple cases, based on both qualitative and quantitative methods is encouraged. Future research avenues can be: (1) geographically varied nature of path dependence, lock-ins and path creation by examining the mechanisms, multi-scalar institutional contexts and global production network dynamics involved; (2) Linking institution theories to IEG and REG to understand how institutions and their embedded agentic strategies/actions evolve, and how they logically affect the adaptation and adaptability dynamics by understanding the structure-agency relationships in OICs and vice versa; (3) An in-depth examination of the relationship between historically conditioned varieties of regional/national capitalism, macro social-political structures (versions of political arrangements) and the nature of OIC resilience trajectories and characteristics; (4) engaging more analysis on the relationship between the nature of crises and regional economic resilience mechanism—which is not opening a promising window of opportunity to reconceptualize the conventional idea of regional economic resilience that has been contextualized in the 2008 financial crisis (Hu et al., 2022).

References

Barnes, T. J., & Sheppard, E. (2010). 'Nothing includes everything': Towards engaged pluralism in Anglophone economic geography. *Progress in Human Geography, 34*, 193–214.

Birch, K., MacKinnon, D., & Cumbers, A. (2010). Old industrial regions in Europe: A comparative assessment of economic performance. *Regional Studies, 44*(1), 35–53.

Boschma, R. (2015). Towards an evolutionary perspective on regional resilience. *Regional Studies, 49*(5), 733–751.

Boschma, R. (2022). Global value chains from an evolutionary economic geography perspective: A research agenda. *Area Development and Policy, 7*(2), 123–146.

Boschma, R., & Capone, G. (2015). Institutions and diversification: Related versus unrelated diversification in a varieties of capitalism framework. *Research Policy, 44*(10), 1902–1914.

Bristow, G., & Healy, A. (2014). Regional resilience: An agency perspective. *Regional Studies, 48*(5), 923–935.

Dawley, S. (2013). Creating new paths? Offshore wind, policy activism, and peripheral region development. *Economic Geography, 90*(1), 91–112.

Dodds, K., Broto, V. C., Detterbeck, K., et al. (2020). The COVID-19 pandemic: Territorial, political and governance dimensions of the crisis. *Territory, Politics, Governance, 8*(3), 289–298.

Economist. (2015). Pollution: The cost of clean air. *The Economist*.

Gereffi, G. (2020). What does the COVID-19 pandemic teach us about global value chains? The case of medical supplies. *Journal of International Business Policy, 3*(3), 287–301.

Glaeser, E. L. (2022). Urban resilience. *Urban Studies, 59*(1), 3–35.

Grabher, G. (1993). The weakness of strong ties: The lock-in of regional development in the Ruhr Area. In G. Grabher (Ed.), *The embedded firm: On the socio-economics of industrial networks* (pp. 255–277). Routledge.

Hassink, R. (2010). Regional resilience: A promising concept to explain differences in regional economic adaptability? *Cambridge Journal of Regions, Economy and Society, 3*(1), 45–58.

Hassink, R., Isaksen, A., & Trippl, M. (2019). Towards a comprehensive understanding of new regional industrial path development. *Regional Studies, 53*(11), 1636–1645.

Hu, X. H. (2014). State-led path creation in China's rustbelt: The case of Fuxin. *Regional Studies, Regional Science, 1*(1), 294–300.
Hu, X. (2017). From coal mining to coal chemicals? Unpacking new path creation in an old industrial region of transitional China. *Growth and Change, 48*(2), 233–245.
Hu, X., & Hassink, R. (2017). Place leadership with Chinese characteristics? A case study of the Zaozhuang coal-mining region in transition. *Regional Studies, 51*(2), 224–234.
Hu, X., Li, L., & Dong, K. (2022). What matters for regional economic resilience amid COVID-19? Evidence from cities in Northeast China. *Cities, 120*, 103440.
Hu, X., & Yang, C. (2018). Building a role model for rust belt cities? Fuxin's economic revitalization in question. *Cities, 72*, 245–251.
Hu, X., & Yang, C. (2019). Institutional change and divergent economic resilience: Path development of two resource-depleted cities in China. *Urban Studies, 56*(16), 3466–3485.
Hu, X., Zhu, S., & Hassink, R. (2020). Beyond the "evolutionary approach": A critical review and paradigmatic reflections on the restructuring of old industrial areas. *Geographical Research., 39*(5), 1028–1044. (In Chinese).
Hudson, R. (2005). Rethinking change in old industrial regions: Reflecting on the experiences of North East England. *Environment and Planning A, 37*(4), 581–596.
MacKinnon, D., Dawley, S., Pike, A., & Cumbers, A. (2019). Rethinking path creation: A geographical political economy approach. *Economic Geography, 95*(2), 113–135.
Martin, R. (2010). Rethinking regional path dependence: Beyond lock-in to evolution. *Economic Geography, 86*(1), 1–27.
Martin, R., & Sunley, P. (2015). On the notion of regional economic resilience: Conceptualization and explanation. *Journal of Economic Geography, 15*, 1–42.
Morgan, K. J. (2012). Path dependence and the state: The politics of novelty in old industrial regions. In P. Cooke (Ed.), *Re-framing regional development: Evolution, innovation and transition, regions and cities* (pp. 318–340). Routledge.
Pike, A., Dawley, S., & Tomaney, J. (2010). Resilience, adaptation and adaptability. *Cambridge Journal of Regions, Economy and Society, 3*(1), 59–70.
Raco, M., & Street, E. (2012). Resilience planning, economic change and the politics of post-recession development in London and Hong Kong. *Urban Studies, 49*(5), 1065–1087.
Strambach, S., & Halkier, H. (2013). Reconceptualising change: path dependency, path plasticity and knowledge combination. *Zeitschrift fuer Wirtschaftsgeographie, 57*(1–2), 1–14.
Yeung, H. W. C. (2005). Rethinking relational economic geography. *Transactions of the Institute of British Geographers, 30*(1), 37–51.

GPSR Compliance

The European Union's (EU) General Product Safety Regulation (GPSR) is a set of rules that requires consumer products to be safe and our obligations to ensure this.

If you have any concerns about our products, you can contact us on ProductSafety@springernature.com

In case Publisher is established outside the EU, the EU authorized representative is:

Springer Nature Customer Service Center GmbH
Europaplatz 3
69115 Heidelberg, Germany